The ESSENTIALS® of

Geometry II

Staff of Research and Education Association

This book is a continuation of *"THE ESSENTIALS OF GEOMETRY I"* and begins with Chapter 9. It covers the usual course outline of Geometry II. Earlier/basic topics are covered in *"THE ESSENTIALS OF GEOMETRY I."*

Research & Education Association
61 Ethel Road West
Piscataway, New Jersey 08854

THE ESSENTIALS®
OF GEOMETRY II

Year 2005 Printing

Printed in the United States of America

Library of Congress Control Number 97-69912

International Standard Book Number 0-87891-607-5

B05

What REA's Essentials®
Will Do for You

This book is part of REA's celebrated Essentials® series of review and study guides, relied on by tens of thousands of students over the years for being complete yet concise.

Here you'll find a comprehensive summary of the very material you're most likely to need for exams, not to mention homework—eliminating the need to read and review many pages of textbook and class notes.

This slim volume condenses the vast amount of detail characteristic of the subject matter and summarizes the essentials of the field. The book provides quick access to the important facts, terms, theorems, concepts, and formulas in the field.

It will save you hours of study and preparation time.

This Essentials® book has been prepared by experts in the field and has been carefully reviewed to ensure its accuracy and maximum usefulness. We believe you'll find it a valuable, handy addition to your library.

Larry B. Kling
Chief Editor

CONTENTS

This book is a continuation of *"THE ESSENTIALS OF GEOMETRY I"* and begins with Chapter 9. It covers the usual course outline of Geometry II. Earlier/basic topics are covered in *"THE ESSENTIALS OF GEOMETRY I"*.

CHAPTER 9

CIRCLES

9.1 CIRCLES/CENTRAL AND INSCRIBED ANGLES/ARCS/CHORDS/SECANTS/TANGENTS/SECTORS AND SEGMENTS OF A CIRCLE

Definition 1

A circle is a set of points in the same plane equidistant from a fixed point called its center.

Definition 2

A radius of a circle is a line segment drawn from the center of the circle to any point on the circle.

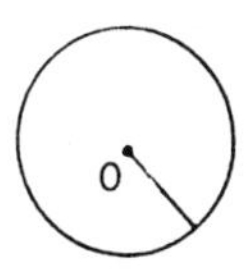

Definition 3

A portion of a circle is called an arc of the circle.

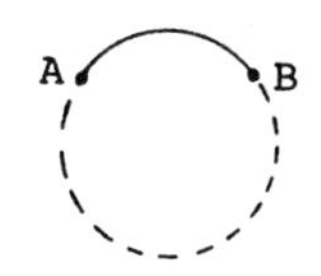

Definition 4

A semicircle is an arc of a circle whose endpoints lie on the extremities of a diameter of the circle.

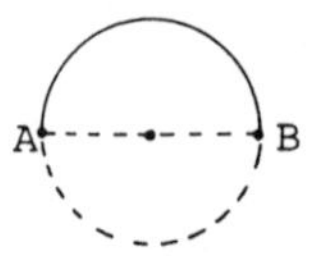

Definition 5

An arc greater than a semicircle is called a major arc.

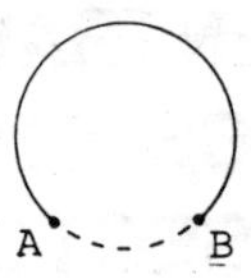

Definition 6

An arc less than a semicircle is called a minor arc.

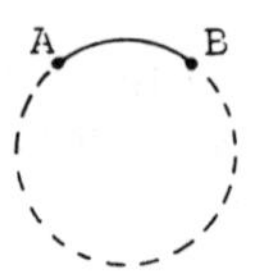

Definition 7

The circumference of a circle is the distance about a circle. The circumference of a circle is given by the formula, $C = 2\pi r$, where r is the radius of the circle.

Definition 8

A sector of a circle is the set of points between two radii and their intercepted arc.

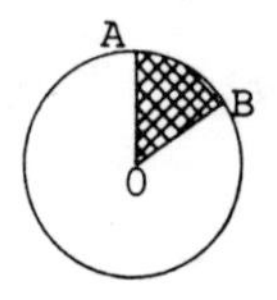

Definition 9

A line that intersects a circle in two points is called a secant.

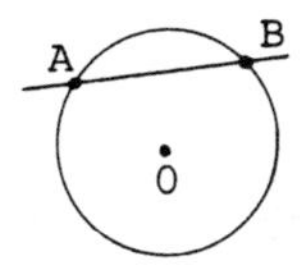

Definition 10

A line segment joining two points on a circle is called a chord of the circle.

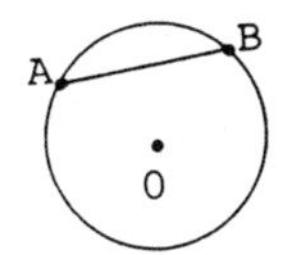

Definition 11

A chord that passes through the center of the circle is called a diameter of the circle.

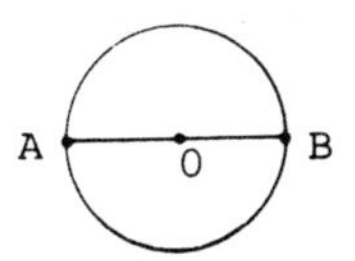

Definition 12

The line passing through the centers of two (or more) circles is called the line of centers.

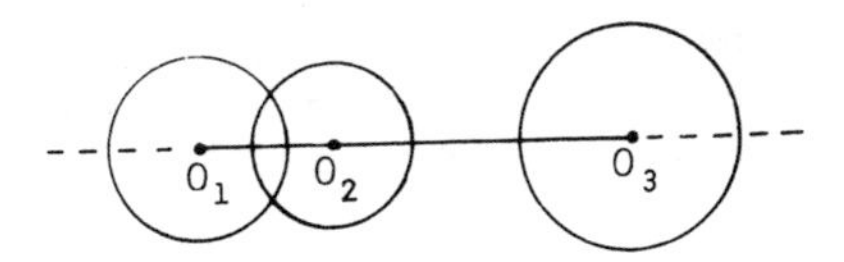

Definition 13

An angle whose vertex is on the circle and whose sides are chords of the circle is called an inscribed angle.

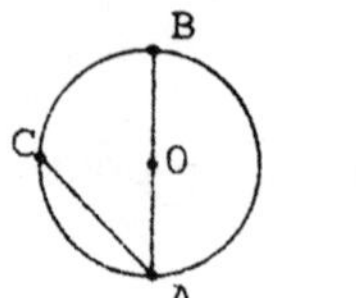

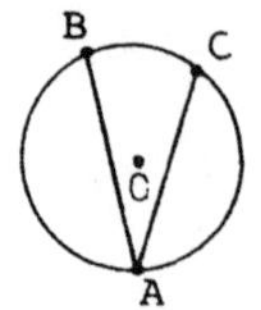

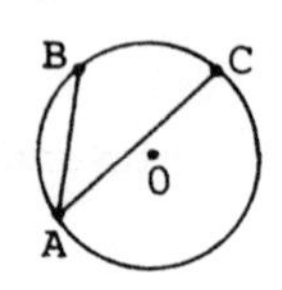

Definition 14

An angle whose vertex is at the center of a circle and whose sides are radii is called a central angle.

The measure of a minor arc is the measure of the central angle that intercepts that arc.

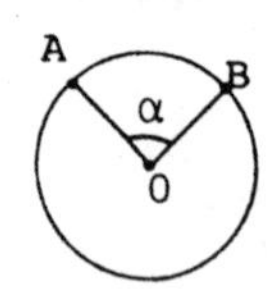

$$m\overset{\frown}{AB} = \alpha = m < AOB$$

Definition 15

The distance from a point P to a given circle is the distance from that point to the point where the circle intersects with a line segment with endpoints at the center of the circle and point P.

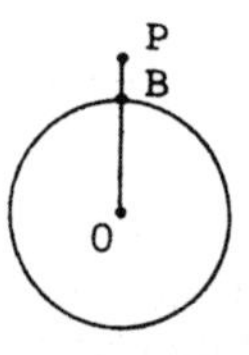

The distance of point P to the diagrammed circle with center O is the line segment PB of line segment PO.

Definition 16

A line that has one and only one point of intersection with a circle is called a tangent to that circle, while their common point is called a point of tangency.

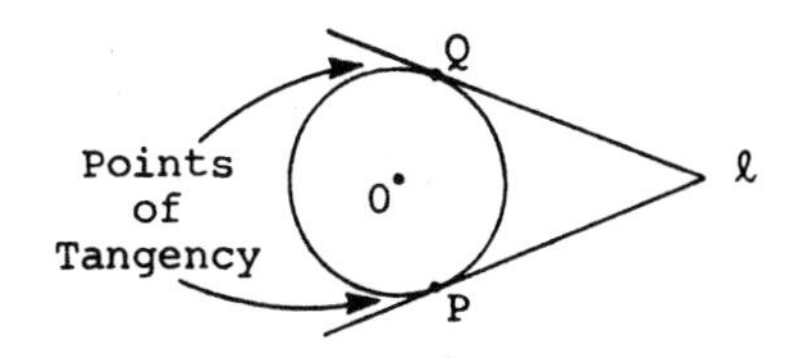

Definition 17

The shaded region below is the union of a circle and its interior.

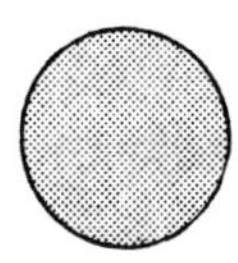

Definition 18

A segment of a circle is a region bounded by an arc of a given circle and the chord containing the endpoints of that arc.

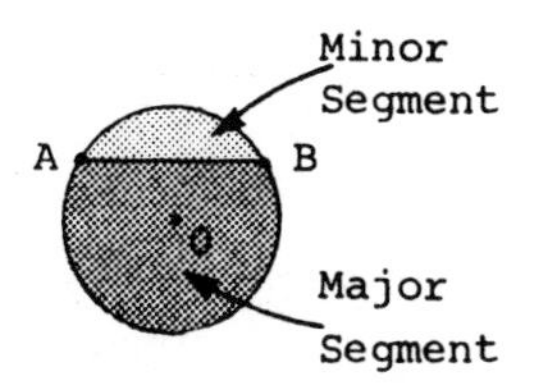

Definition 19

Congruent circles are circles whose radii are congruent.

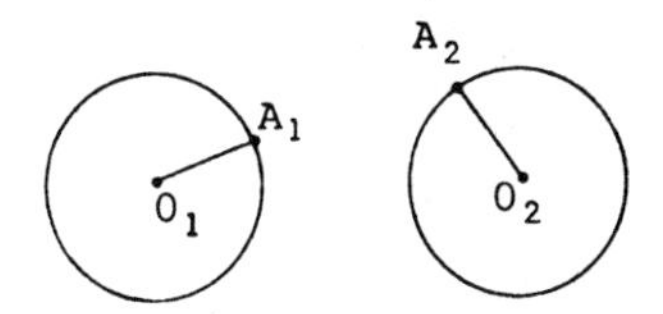

If $O_1A_1 \cong O_2A_2$, then $O_1 \cong O_2$.

Definition 20

The measure of a semicircle is 180°.

Definition 21

A quadrant is an arc whose measure is 90°.

Definition 22

The measure of a major arc is 360° minus the measure of the minor arc formed with the same endpoints.

Definition 23

Congruent arcs are arcs that have equal degree measures and equal lengths.

Definition 24

The midpoint of an arc is the point that divides the arc into two congruent arcs.

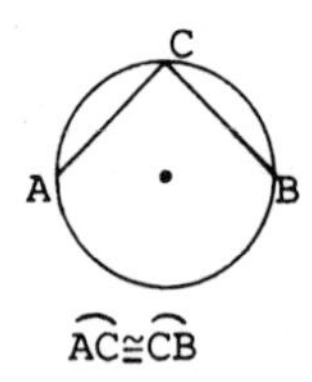

Definition 25

A circle is inscribed in a polygon if all the sides of the polygon are tangential to the circle.

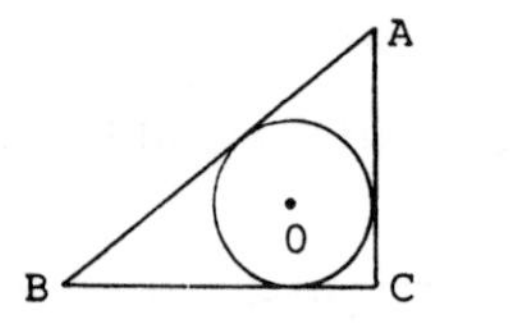

Definition 26

The length of a tangent from an external point to a given circle is the length of the line segment with endpoints at external point and the point of tangency.

Definition 27

A line which is tangent to two circles is called the common tangent to the circles.

Definition 28

A common internal tangent to two circles is a line which is tangent to both circles and intersects their line of centers at a point between the two centers.

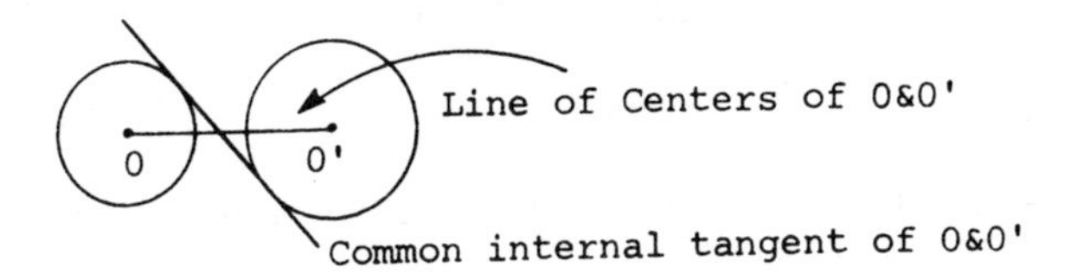

Definition 29

A common tangent of two circles which does not intersect the line of centers of the two circles at a point between the two centers is called the common external tangent of the given circles.

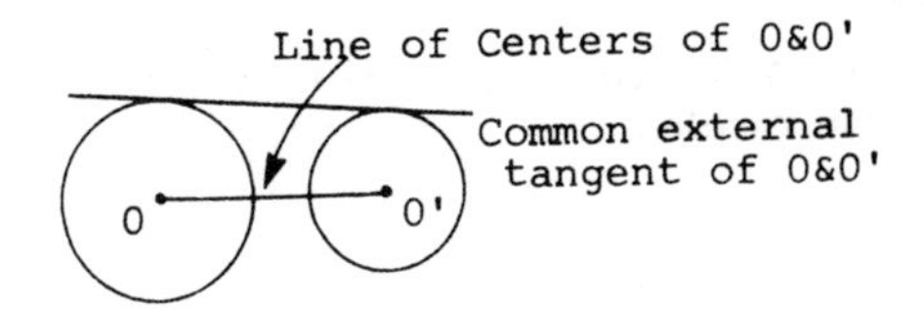

Definition 30

Circles that are in a plane and tangent to the same line at the same point are tangent circles.

If two tangent circles lie on the same side of the common tangent they are internally tangent; otherwise, they are externally tangent.

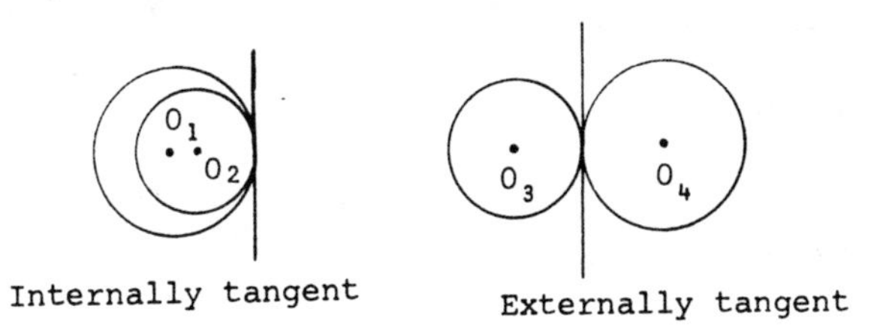

Definition 31

An angle is inscribed in an arc of a circle if its vertex lies on the arc and its sides are chords which join the vertex and ends of the arc.

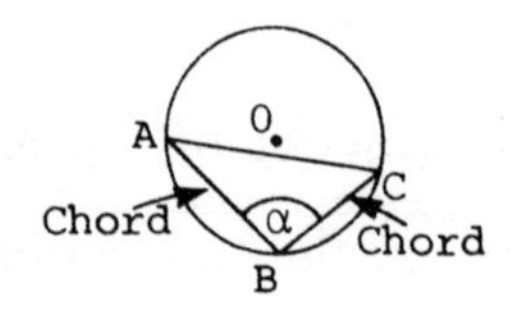

$\angle\alpha$ inscribed in $\widehat{AC}$

Definition 32

A circumscribed circle is a circle passing through all the vertices of a polygon.

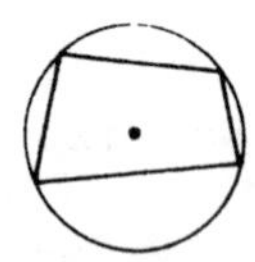

Definition 33

Circles that have the same center and unequal radii are called concentric circles.

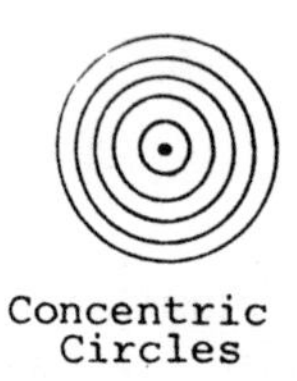

Concentric Circles

Postulate 1

Two circles are congruent if and only if their radii (or diameters) are equal.

Postulate 2

A line drawn from a center of a circle to a point of tangency is perpendicular to the tangent passing through the point of tangency.

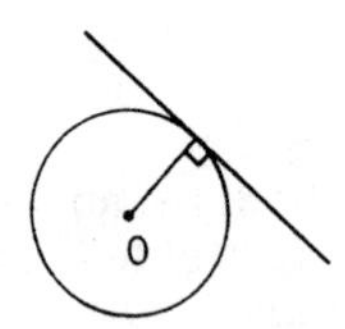

Postulate 3

In the same circle or in two congruent circles, arcs of same degree have equal lengths.

Postulate 4

One and only one circle can be drawn in a plane with a given point as its center and a given line segment as its radius.

Postulate 5

Except for the point of tangency, all points on the tangent line of a circle lie outside the circle.

Theorem 1

The length of a diameter is twice the length of a radius.

Theorem 2

In a circle, parallel lines intercept equal arcs.

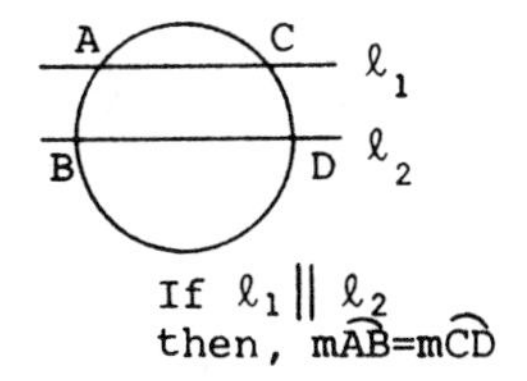

Theorem 3

In the same circle or congruent circles equal arcs have equal chords.

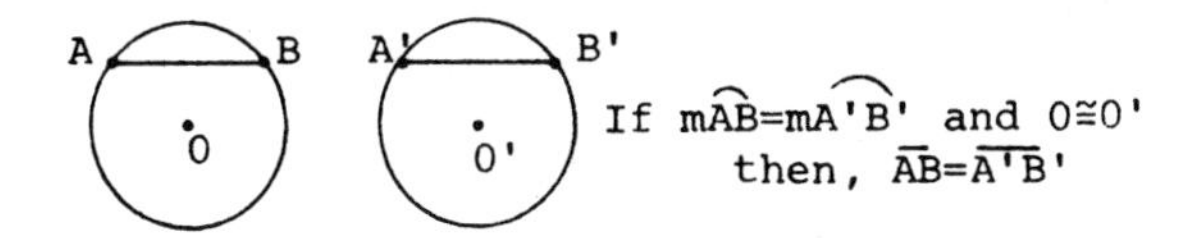

Theorem 4

In the same circle or congruent circles, equal chords are equidistant from the center.

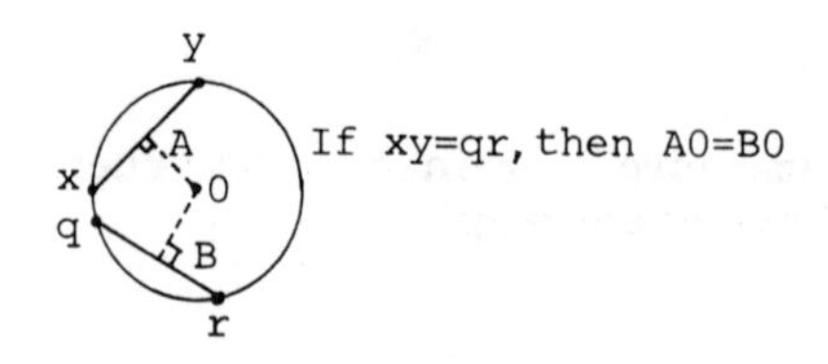

Theorem 5

In the same circle or congruent circles, chords equidistant from the center are equal.

Theorem 6

If two chords intersect within a circle, the product of the segments of one chord is equal to the product of the segments of the other chord.

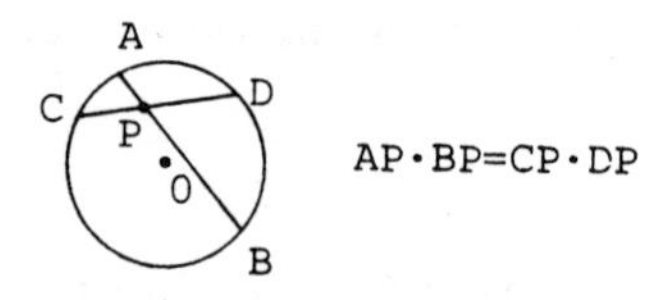

Theorem 7

In the same circle or in congruent circles, equal chords have equal arcs.

Theorem 8

In the same circle or in congruent circles, two arcs which contain the same number of arc degrees are equal.

Theorem 9

If two secants are drawn to a circle from a point outside the circle, the products of the secants and their external segments are equal.

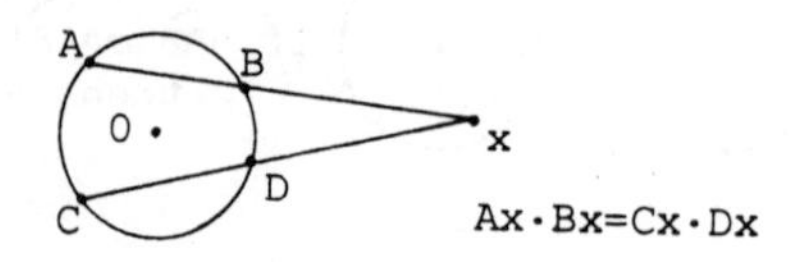

Theorem 10

The circumference of a circle contains 360°.

Theorem 11

In a circle, the length of an arc which contains $n°$ or whose central angle contains $n°$, is given by the formula:

$$\text{length of arc} = \left(\begin{array}{l}\text{circumference}\\ \text{of a circle}\end{array}\right) \times \frac{n°}{360°}$$

$$= 2\pi r \times \frac{n°}{360°}$$

$$= \pi D \ \frac{n°}{360°}$$

where r and D are the radius and diameter of the given circle, respectively.

Theorem 12

If two arcs of the same circle, $\overset{\frown}{AB}$ and $\overset{\frown}{BC}$, have only one point in common (point B), and if their union is $\overset{\frown}{AC}$, then $m\overset{\frown}{AB} + m\overset{\frown}{BC} = m\overset{\frown}{AC}$.

A

B O C

$m\overset{\frown}{AB}+m\overset{\frown}{BC}=m\overset{\frown}{AC}$

Theorem 13

If from a point outside a given circle, a tangent and a secant are drawn to the circle, then the length of the tangent will be the mean proportional between the length of the secant and the length of its external segment.

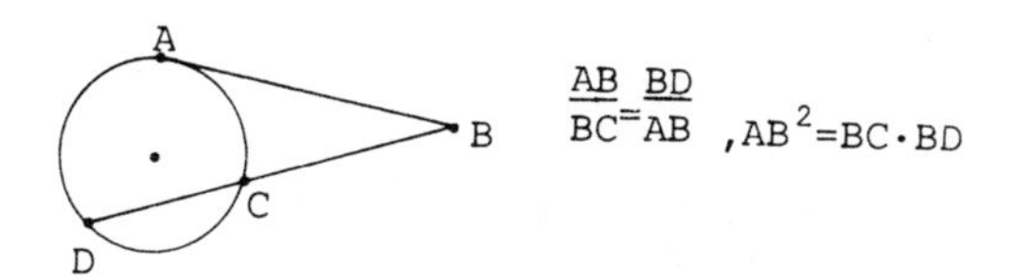

Theorem 14

A diameter divides a circle into two equal parts, if a chord divides a circle into two equal parts, then it is a diameter.

Theorem 15

Radii of the same circle or congruent circles are equal.

Theorem 16

Diameters of the same circle or of congruent circles are equal.

Theorem 17

The measure of an inscribed angle is equal to one-half the measure of its intercepted arc.

Theorem 18

An angle inscribed in a semicircle is a right angle.

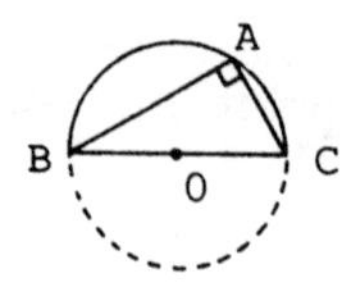

Theorem 19

An angle formed by the intersection of two secants outside a circle is equal to one-half the difference of its intercepted arcs.

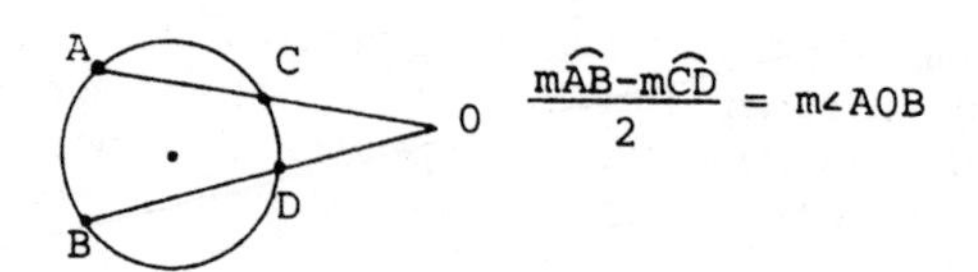

Theorem 20

An angle formed by a tangent and a chord is equal to one-half the measure of the intercepted arc.

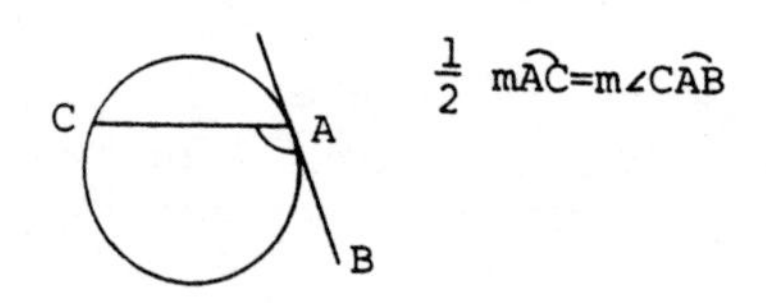

Theorem 21

If two chords intersect within a circle, each angle formed is equal to one-half the sum of its intercepted arc and the intercepted arc of its vertical angle.

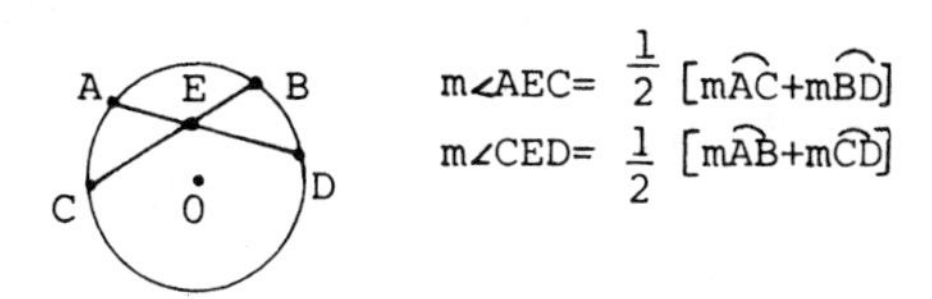

Theorem 22

An angle formed by the intersection of a tangent and a secant outside a circle is equal to one-half the difference of the intercepted arcs.

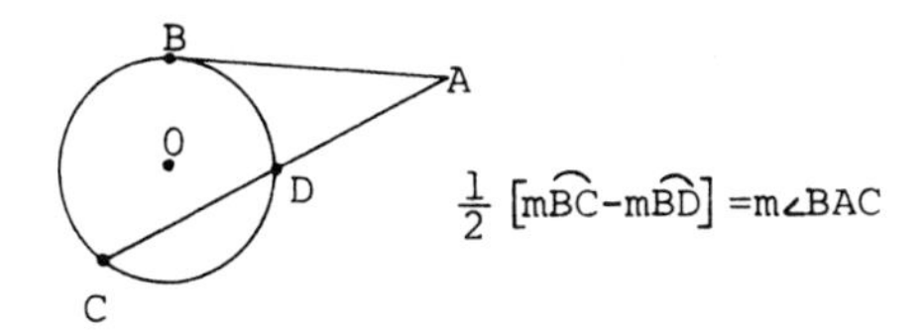

Theorem 23

The measure of an angle formed by two tangents drawn to a circle from an outside point is equal to one-half the difference of the measures of the intercepted arcs.

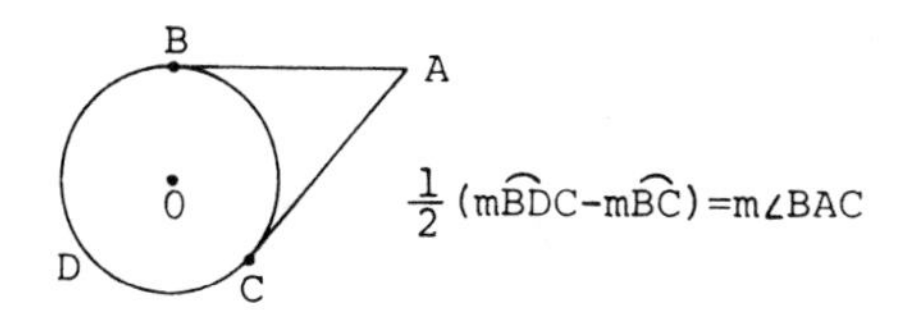

Theorem 24

In the same circle or in congruent circles, central angles of equal measures have equal arcs.

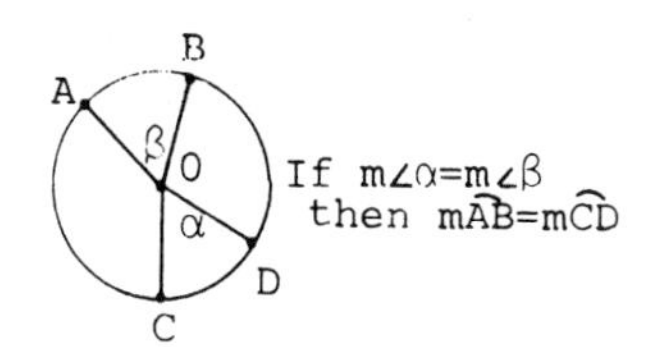

Theorem 25

In the same circle or in congruent circles, equal arcs have central angles whose measures are equal.

Theorem 26

In the same or congruent circles, equal inscribed angles have equal intercepted arcs.

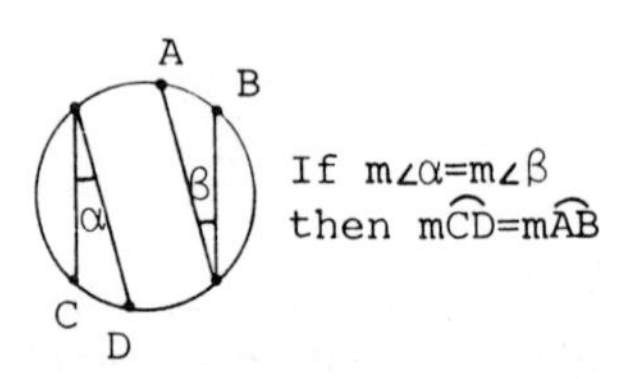

Theorem 27

In the same circle or in congruent circles, inscribed angles which have equal intercepted arcs are equal.

Theorem 28

Angles inscribed in the same or equal arcs have equal measures.

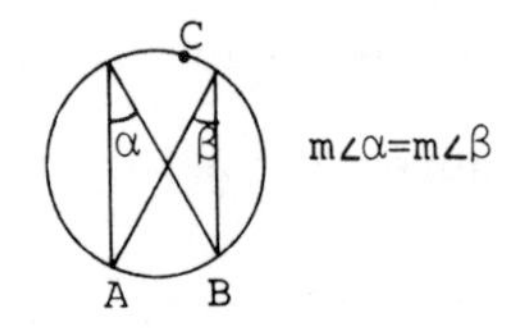

Note: All angles which share endpoints A and B and have a vertex lying in ACB have equal measures.

Theorem 29

Opposite angles of an inscribed quadrilateral are supplementary.

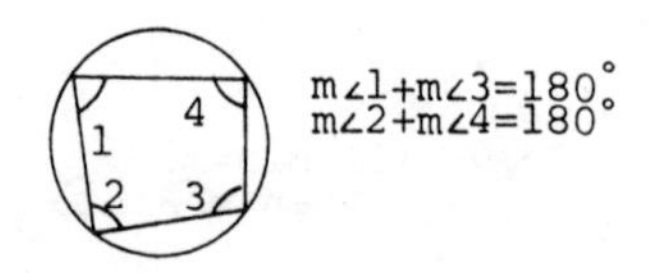

Theorem 30

A) From any point lying outside a given circle, exactly two tangents can be drawn to that circle.

B) From any point lying on the circumference of a given circle, exactly one tangent can be drawn to that circle.

C) From any point lying inside a given circle, no tangent line can be drawn to that circle.

Theorem 31

A diameter perpendicular to a chord of a circle bisects the chord and its arcs.

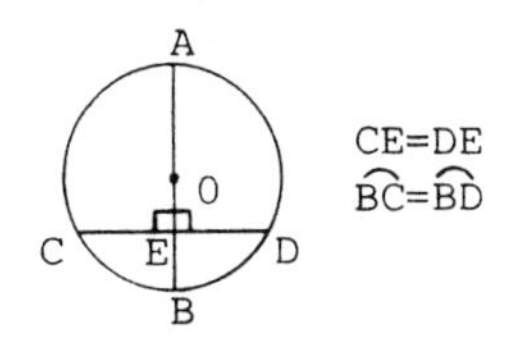

Theorem 32

A tangent line to a given circle is perpendicular to the radius drawn to the point of tangency.

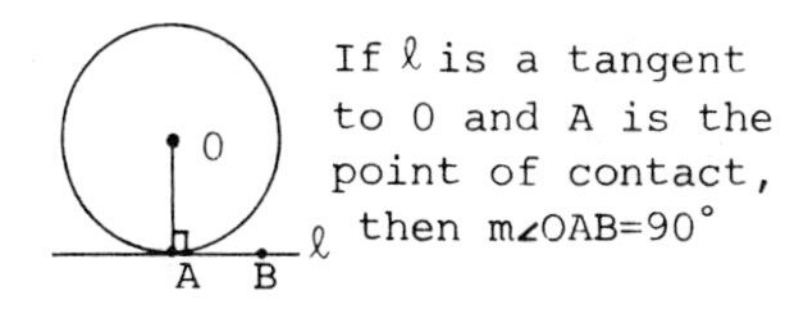

Theorem 33

If a line is perpendicular to a radius of a circle at its outer endpoint, then the line is a tangent to the circle.

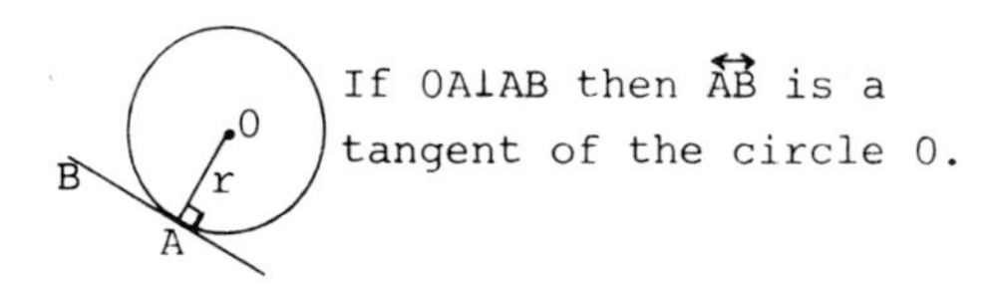

Theorem 34

If two tangents are drawn to a circle from an external point, these tangents are equal in length.

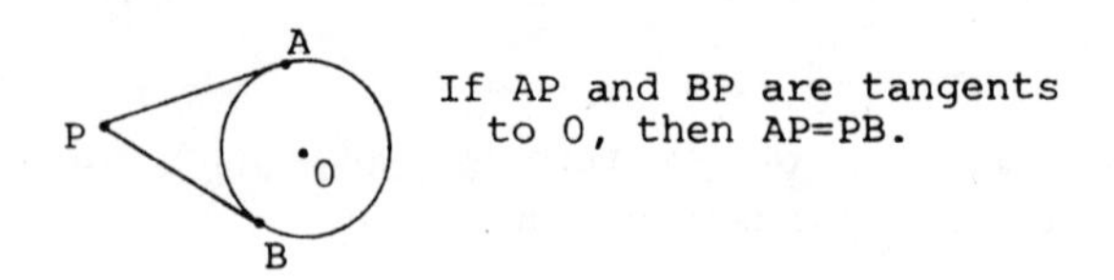

If AP and BP are tangents to O, then AP=PB.

Theorem 35

If two circles intersect in two points, then their line of centers is the perpendicular bisector of their common chord.

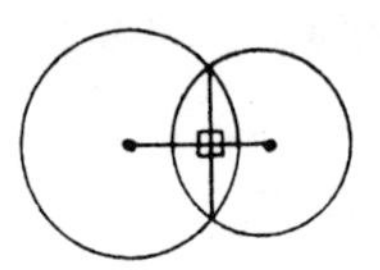

Theorem 36

If two circles are tangential, then their line of centers (extended if necessary) passes through the point of contact and is perpendicular to their common tangent.

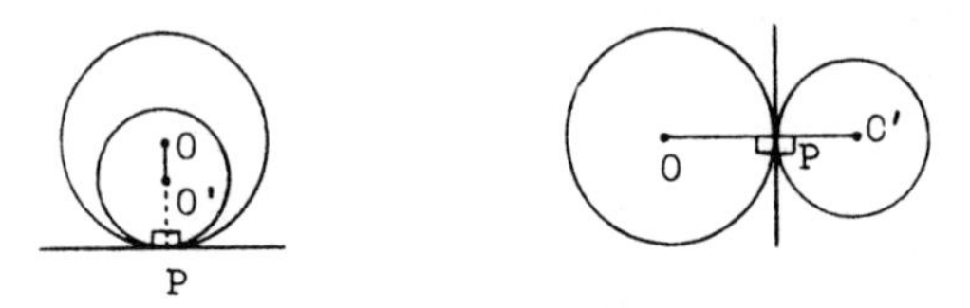

Theorem 37

A circle can be inscribed in any regular polygon.

Theorem 38

A circle can be circumscribed about any regular polygon.

Corollary 1

The perpendicular bisector of a chord of a given circle passes through the center of the circle.

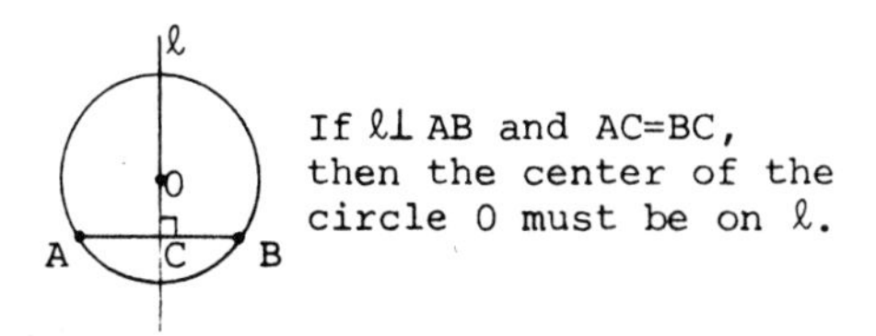

Corollary 2

If two tangents are drawn to a given circle from an external point, then the line passing through that external point and the center of the circle is the angle bisector of the angle formed by the tangents.

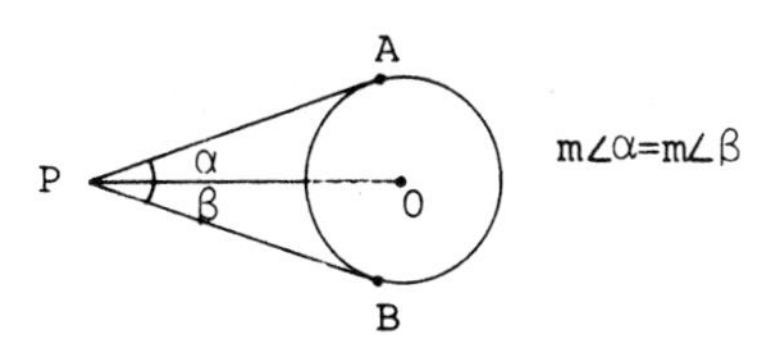

Corollary 3

If a line passes through the center of a given circle, and bisects a chord that is not a diameter, then it is perpendicular to the chord.

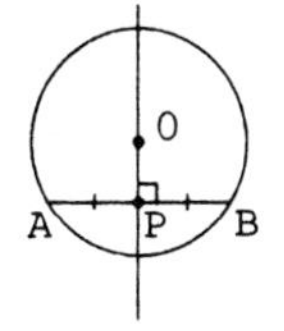

If $\overleftrightarrow{OP}$ bisects chord AB then $\overleftrightarrow{OP} \perp \overline{AB}$

Corollary 4

If a line is perpendicular to a tangent of a circle at the point of tangency, it passes through the center of the circle.

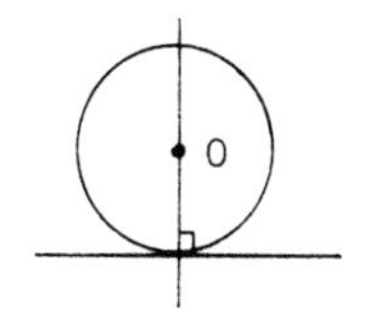

9.2 AREAS

Definition

The area of a circle is given by: $A = \pi r^2$, where r is the radius of the circle.

Axiom

The area of a sector of a circle divided by the area of the entire circle is equal to the measure of its central angle divided by 360°.

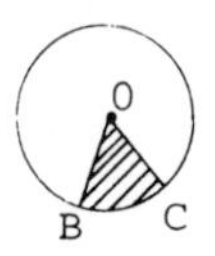

$$\frac{\text{Area of shaded area OBC}}{\text{Area of circle center O}} = \frac{\angle BOC°}{360°}$$

Theorem 1

The area of a circle is equal to one-half the product of its radius and circumference:

$$A = \tfrac{1}{2}rc,$$

where r and c are the radius and circumference of the circle, respectively.

Theorem 2

The area of a circle is equal to $\frac{1}{4}\pi$ times the square of the length of the diameter:

$$A = \tfrac{1}{4}\pi d^2.$$

Corollary 1

The ratio of the areas of two circles equals the square of their ratio of similitude.

Circle with center O_1 and radius r_1

Circle with center O_2 and radius r_2

$$\frac{\pi r_1^2}{\pi r_2^2} = \left(\frac{r_1}{r_2}\right)^2$$

Corollary 2

The ratio of the circumference of two circles is equal to the ratio of the lengths of their radii or diameters. Thus:

$$\frac{c_1}{c_2} = \frac{2\pi r_1}{2\pi r_2} = \frac{r^1}{r^2}, \quad \frac{c_1}{c_2} = \frac{\pi d_1}{\pi d_2} = \frac{d_1}{d_2}.$$

Theorem 3

The area of a minor segment of a circle is equal to the area of its sector minus the area of the triangle formed by its radii and the chord.

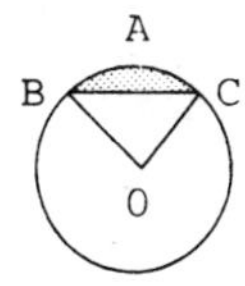

The shaded region represents the area of a minor segment of circle O.

$$\left(\begin{array}{c}\text{Area of the}\\ \text{shaded region}\end{array}\right) = \left(\begin{array}{c}\text{Area of the}\\ \text{sector BOC}\end{array}\right) - \left(\text{Area of } \Delta BOC\right)$$

9.3 INEQUALITIES INVOLVING CIRCLES

Theorem 1

In the same circle or in equal circles, the greater of two central angles will intercept the greater arc.

If $m\angle\beta > m\angle\alpha$, then $m\overset{\frown}{BC} > m\overset{\frown}{AB}$.

Theorem 2

In the same circle or in equal circles, the greater of

two arcs will be intercepted by the greater of the central angles.

Theorem 3

In the same circle or equal circles, the greater of two chords intercepts the greater minor arc.

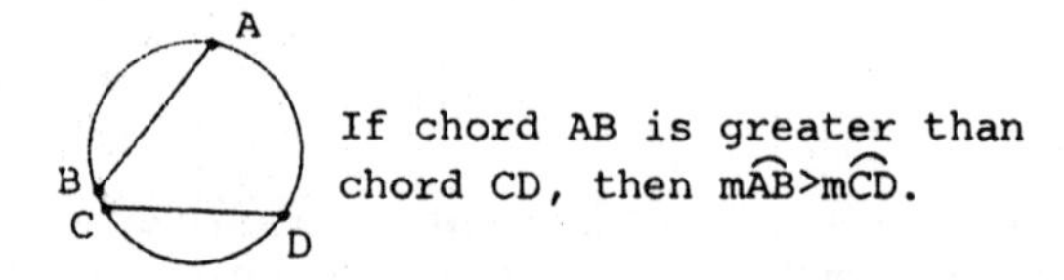

Theorem 4

In the same circle or equal circles, the greater of two minor arcs has the greater chord.

Theorem 5

If two unequal chords form an inscribed angle within a circle, the shorter chord is the farther from the center of the circle.

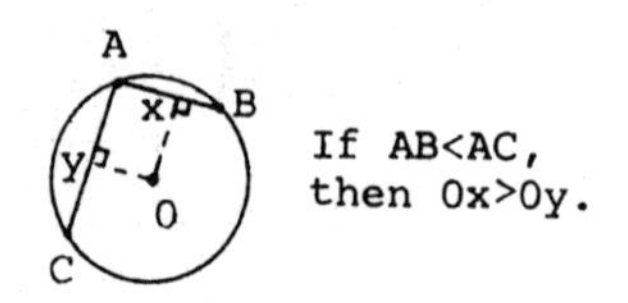

Theorem 6

In the same circle or in equal circles, if two chords are not equal, then they are at unequal distances from the center of the circle, and the greater chord is nearer to the center.

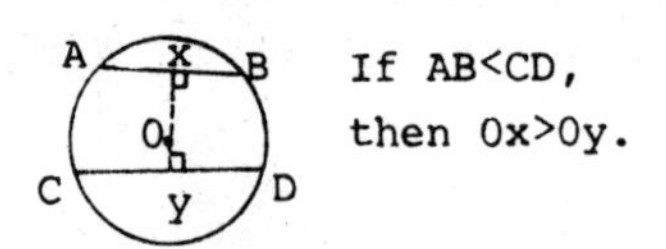

Theorem 7

In the same circle or in equal circles, if two chords are at unequal distances from the center of the circle, then they are not congruent, and the chord nearer the center is greater.

9.4 LAW OF SINES

Definition 1

The center of the inscribed circle of a triangle is called the incenter of the triangle.

Definition 2

The center of the circumscribed circle of a triangle is called the circumcenter.

Theorem 1

The angle bisectors of a triangle are concurrent at the incenter of the circle which can be inscribed in that triangle.

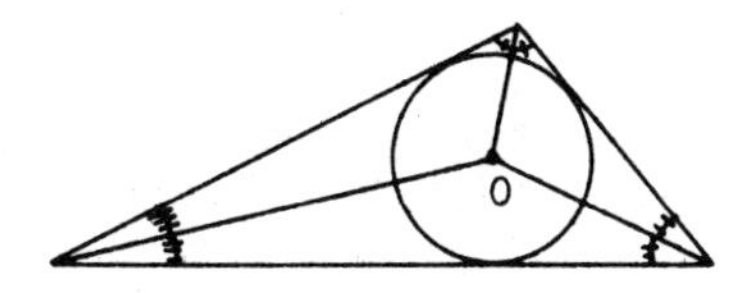

Theorem 2

The perpendicular bisectors of the sides of a triangle are concurrent at the circumcenter of the circle that can be circumscribed about that triangle.

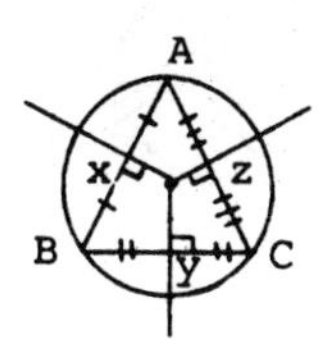

Law of Sines

For a triangle ΔABC and its circumscribed circle of radius

R:

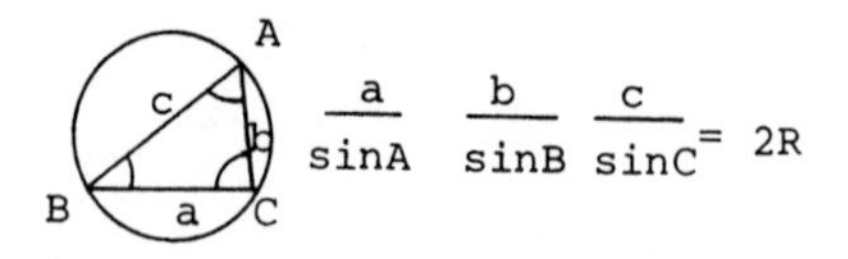

$$\frac{a}{\sin A} = \frac{b}{\sin B} = \frac{c}{\sin C} = 2R$$

Theorem 1

The length of the radius of a circle circumscribed about an equilateral triangle is given by the formula:

$$r = \frac{2}{3}h,$$

where h is the length of the altitude of the triangle.

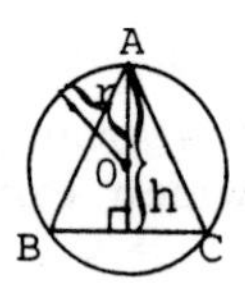

Theorem 2

The length of the radius of a circle inscribed in an equaliteral triangle is given by the formula:

$$r = \frac{1}{3}h,$$

where h is the length of the altitude of the triangle.

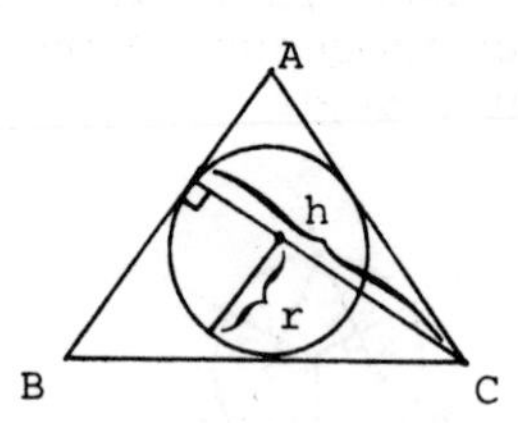

Theorem 3

The length of the radius of a circle circumscribed about an equilateral triangle is twice the length of the radius of the circle inscribed in the same triangle.

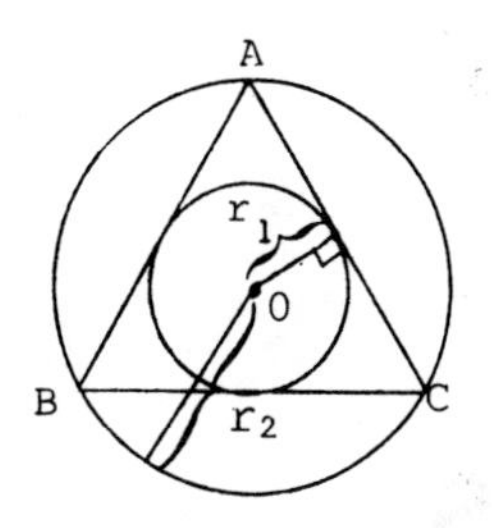

Given: Equilateral ΔABC with inscribed circle center 0 radius r_1 and circumscribed circle center 0 radius r_2.

Hence: $r_2 = 2r_1$.

CHAPTER 10

POLYGONS

10.1 REGULAR POLYGONS

Definition 1

A polygon is a figure with the same number of sides as angles.

Definition 2

An equilateral polygon is a polygon all of whose sides are of equal measure.

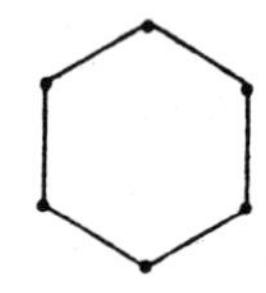

Definition 3

An equiangular polygon is a polygon all of whose angles are of equal measure.

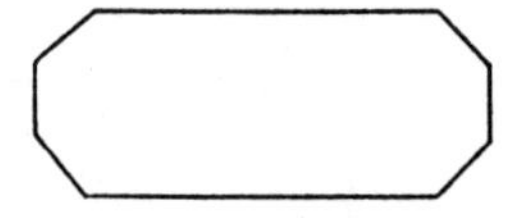

Definition 4

A regular polygon is a polygon that is both equilateral and equiangular.

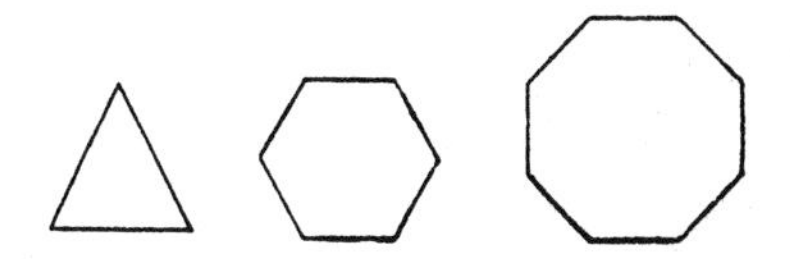

Definition 5

The center of a regular polygon is the center of the circle inscribed within the polygon.

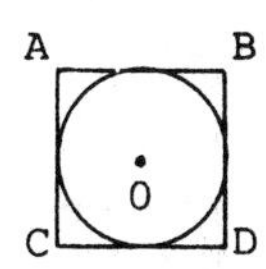

Definition 6

The radius of a regular polygon is a line segment drawn from the center of the polygon to one of its vertices. It is also the radius of the circumscribed circle of the polygon.

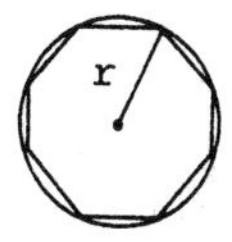

Definition 7

A central angle of a regular polygon is the angle formed by radii drawn to two consecutive vertices.

Definition 8

An apothem of a regular polygon is a radius of its inscribed circle drawn to one of its sides.

Definition 9

A convex polygon is one whose interior angles measure less than 180°.

Definition 10

A concave polygon is one which has at least one interior angle that measures more than 180°.

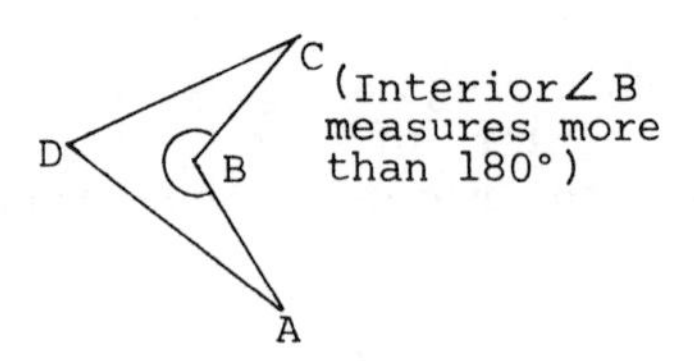

Postulate

When the number of sides of a regular polygon inscribed in a circle increases without bound, then the length of the apothem of the polygon approaches the length of the radius of the circle, and so may be used as an approximation of the length of the radius of the circle. Furthermore, the area of the regular polygon approaches the area of the circle and so can be used as an approximation of the area of the circle.

Theorem 1

The center of a circle that is circumscribed about a regular polygon is the center of the circle that is inscribed within the regular polygon.

Theorem 2

If a circle is divided into n equal arcs $(n > 2)$, the chords of the arcs will form a regular polygon.

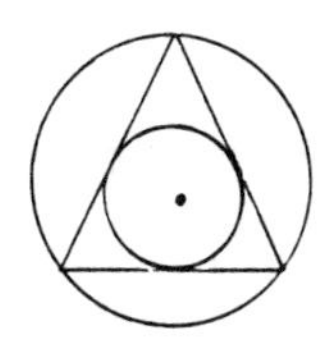

Theorem 3

If a circle is divided into n equal arcs and tangents are drawn to the circle at the endpoints of these arcs, the figure formed by the tangents will be a regular polygon ($n > 2$).

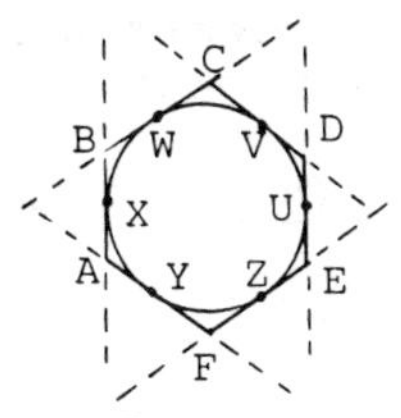

$$\widehat{UV} = \widehat{VW} = \widehat{WX} = \widehat{XY} = \widehat{YZ} = \widehat{ZU}$$

AB, BC, CD, DE, EF and FA are all tangent lines to the circle with respective points of tangency U,V,W,X,Y and Z as shown. Therefore, the polygon ABCDEF formed from the points of intersection of the tangent lines is a regular polygon.

Theorem 4

An apothem of a regular polygon is also the perpendicular bisector of its respective side.

Theorem 5

A central angle of a regular polygon is given by the formula, $x = \frac{2}{n} \times 180°$, where n is the number of sides of the polygon.

Given that ABCDE is a regular polygon with $n = 5$ sides inscribed in circle O, each central angle is equal to $x = \frac{2}{5} \times 180° = 72°$.

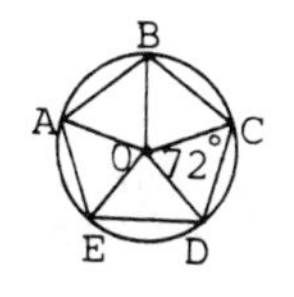

Theorem 6

A radius of a regular polygon bisects the angle of the vertex to which it is drawn.

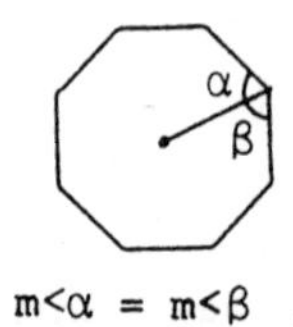

$m\angle\alpha = m\angle\beta$

Theorem 7

The measure of each interior angle of a regular polygon of n sides is given by the formula:

$$\alpha = \frac{(n - 2)}{n} \; 180°.$$

Theorem 8

The measure of each exterior angle of a regular polygon of n sides is determined by:

$$\frac{360°}{n}$$

Theorem 9

The sum of the interior angles of a polygon is given by the formula $S = (n - 2) \cdot 180°$, where n is the number of sides of the polygon.

Theorem 10

The sum of the exterior angles of a polygon, using one angle for each vertex, is always 360°.

Theorem 11

The perimeter of a regular polygon of N sides which is inscribed in a circle of radius R is given by the formula:

$$P = 2NR \sin \frac{180°}{N}$$

Theorem 12

The perimeter of a regular polygon of n sides which is circumscribed about a circle of radius R is represented by:

$$P = 2nR \tan \frac{180°}{n}$$

Theorem 13

A regular polygon of n sides with each side of length A has a perimeter equal to n times a ($P = na$).

Theorem 14

An equilateral polygon inscribed in a circle is a regular polygon.

Theorem 15

The radii of a regular polygon are congruent.

Theorem 16

The apothems of a regular polygon are congruent.

Theorem 17

All of the sides of a circumscribed polygon are tangential to the circle.

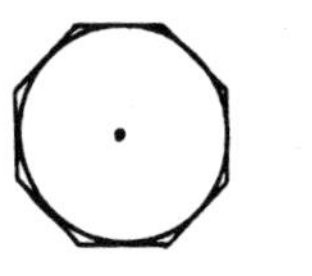

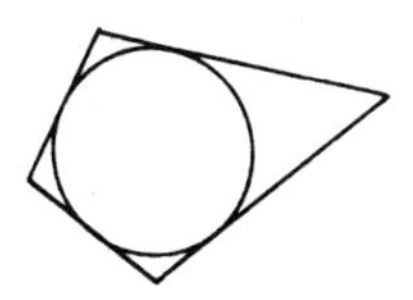

Theorem 18

All of the sides of an inscribed polygon are chords of the circle.

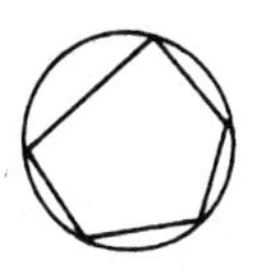

10.2 AREAS OF POLYGONS AND RATIOS OF LINES

Definition 1

Polygons having the same area are called equivalent polygons.

Definition 2

The term polygonal region refers to the union of the polygon and its interior.

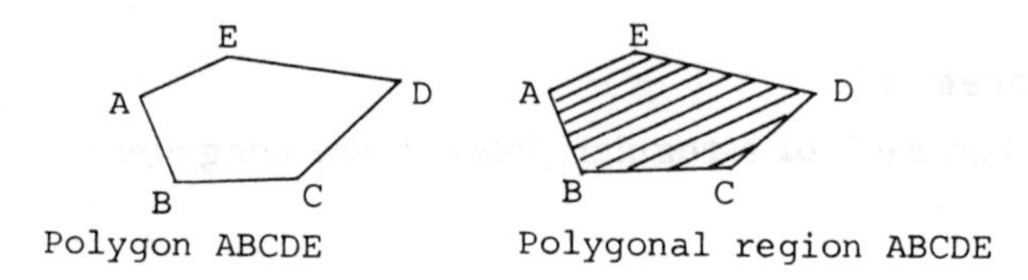

Polygon ABCDE Polygonal region ABCDE

Definition 3

The area of a polygonal region is the maximum unique number which the area can contain.

Theorem 1

The area of a regular polygon of n sides, each side of length S, is equal to the product of one-half the polygon's apothem and its perimeter:

$$\text{Area} = \frac{1}{2} a \cdot nS.$$

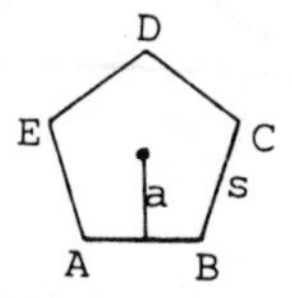

Figure ABCDE is a regular polygon with five sides of length & apothem of length a. The area is thus $\frac{(5 \times 5) \times a}{2}$

Theorem 2

The area of a parallelogram is given by the formula A = bh where b is the length of a base and h is the corresponding height of the parallelogram. The side to which an altitude is drawn is called a base. The length of an altitude is called a height.

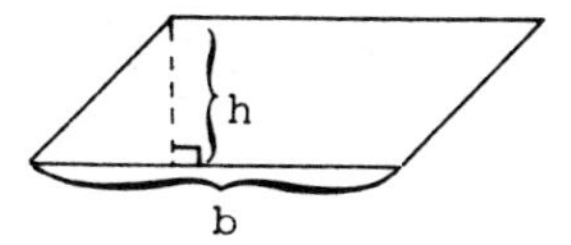

Theorem 3

The area of a square is the square of the length of one side of the square.

$A = s^2$ s s

Theorem 4

The area of a trapezoid is given by the formula $A = \frac{1}{2}h(b_1 + b_2)$, where h is the height and b_1 and b_2 are the lengths of the bases of the trapezoid.

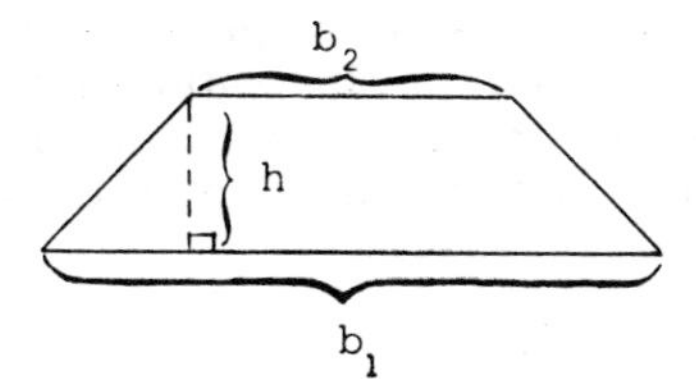

Theorem 5

The area of a rhombus is equal to one-half the product of its diagonals.

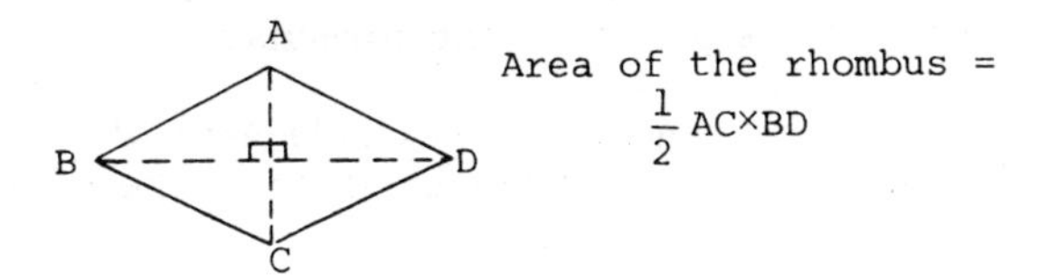

Theorem 6

The area of a regular polygon of n sides inscribed in a circle of radius R, is given by the formula:

$$A = \frac{nr^2}{2} \sin \frac{360°}{n}$$

$$\text{Area} = \frac{5r^2}{2} \cdot \sin \frac{360°}{5} = \frac{5}{2} r^2 \sin 72°.$$

Theorem 7

The area of a regular polygon of n sides which is circumscribed about a circle having radius of length r is given by the formula:

$$A = nr^2 \tan \frac{180°}{n}$$

$$\text{Area} = 5r^2 \tan \frac{180°}{5},$$
$$= 5r^2 \tan 36°.$$

Theorem 8

The ratio of the areas of regular polygons of the same number of sides is equal to the ratio of the squares of the lengths of their sides or the squares of the lengths of their radii, or the squares of the lengths of their apothems.

Theorem 9

Regular polygons with the same number of sides are similar.

Theorem 10

If a regular polygon is inscribed in a circle, then the area of each segment is given by the formula:

$$\text{Area of each segment} = \frac{\text{Area of the circle-area of the polygon}}{\text{Number of sides of the polygon}}$$

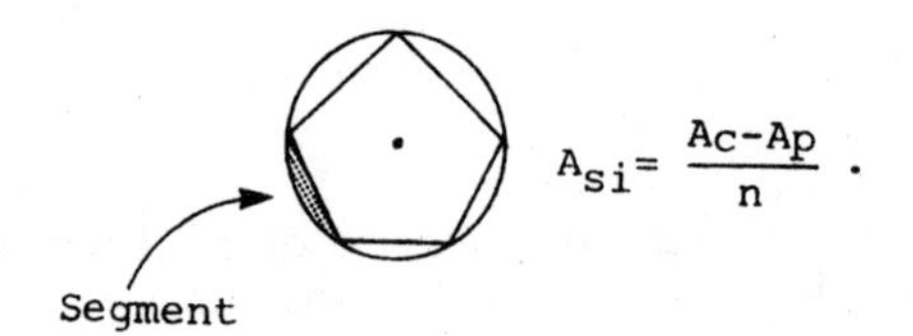

$$A_{si} = \frac{Ac - Ap}{n}.$$

CHAPTER 11

COORDINATE GEOMETRY

11.1 INTRODUCTION, DISTANCE AND MIDDLE POINT FORMULAS

Definition

Coordinate geometry refers to the study of geometric figures using algebraic principles.

Postulate:

The points on a straight line can be placed in a one-to-one correspondence with real numbers such that for every point of the line there corresponds a unique real number, and for every real number there corresponds a unique point of the line.

Definition 1

A number scale is a straight line on which distances from a point are numbered in equal units, positively in one direction and negatively in the other direction. The origin is the zero point from which distances are measured.

Definition 2

The Cartesian product of a set X and a set Y is the set of all ordered pairs (x,y), where x belongs to X and y belongs to Y.

Definition 3

The graph of R × R is called the Cartesian coordinate plane (where R is the set of real numbers). The graph consists of a pair of perpendicular lines called coordinate axes. The vertical axis is the y-axis and the horizontal axis is the x-axis. The point of intersection of these two axes is called the origin; it is the zero point of both axes. Furthermore, points to the right of the origin on the x-axis and above the origin on the y-axis represent positive real numbers. Points to the left of the origin on the x-axis or below the origin on the y-axis represent negative real numbers. Each element of the set R × R is represented by a point on the Cartesian coordinate plane.

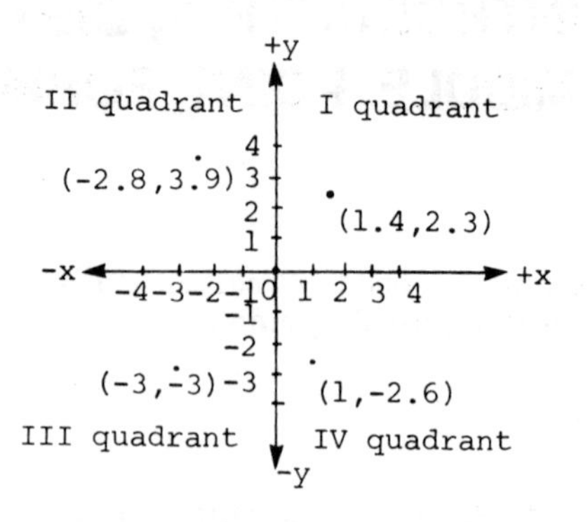

The abscissa is the coordinate of the projection of a given point onto the x-axis, and the ordinate is the coordinate of the projection of a given point onto the y-axis.

The four regions marked off by the coordinate axes are, in counter-clockwise sequence from the top right, described as the first, second, third, and fourth quadrants. The first quadrant contains all points with two positive coordinates.

Definition 4

The distance between any two points on a number scale is the absolute value of the difference between the corresponding number.

Definition 5

Directed distance of a horizontal line segment from one

point to a second is the x-coordinate of the second minus the x-coordinate of the first.

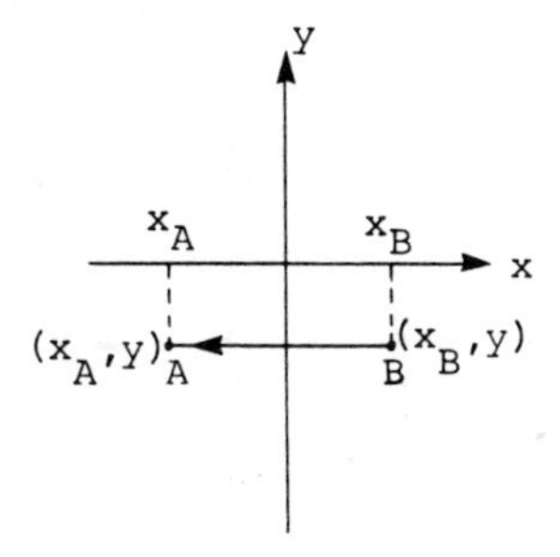

The directed distance of line segment BA from B to A is found by:

$$d = x_A - x_B.$$

Definition 6

Directed distance of a vertical line segment from one point to a second is the y-coordinate of the second minus the y-coordinate of the first.

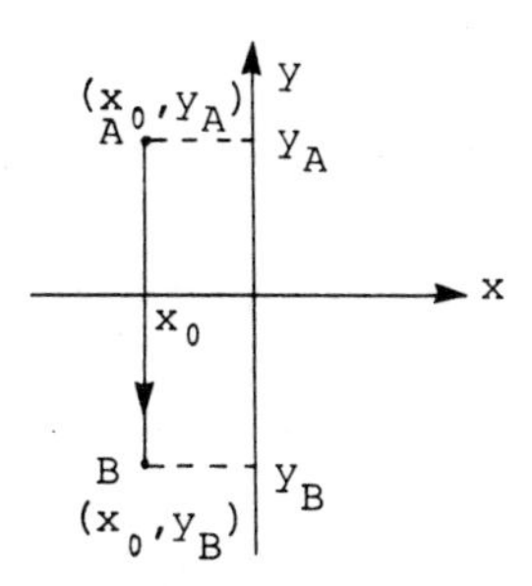

The directed distance from point A to point B of the vertical line segment AB is found by:

$$d = Y_B - Y_A.$$

Theorem 1

For any two points A and B with coordinates (X_A, Y_A) and (X_B, Y_B), respectively, the distance between A and B is represented by:

$$AB = \sqrt{(X_A - X_B)^2 + (Y_A - Y_B)^2}$$

This is commonly known as the distance formula.

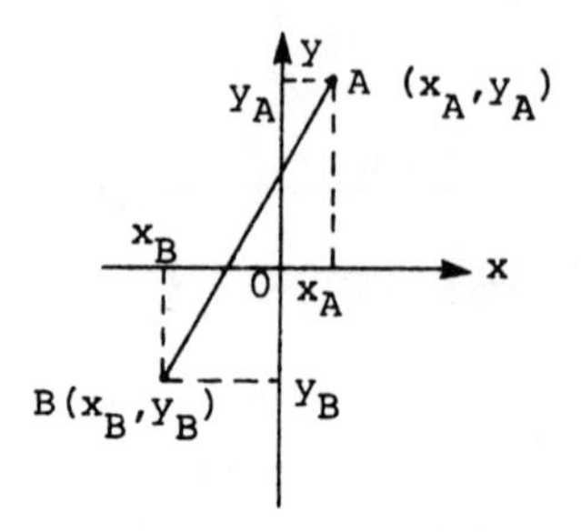

Theorem 2

Given a line segment with endpoints $A(X_A, Y_A)$ and $B(X_B, Y_B)$, the coordinates of the midpoint of this line segment are $M(X_m, Y_m)$:

$$X_m = \frac{X_A + X_B}{2}, \qquad Y_m = \frac{Y_A + Y_B}{2}$$

This is commonly known as the Midpoint Formula.

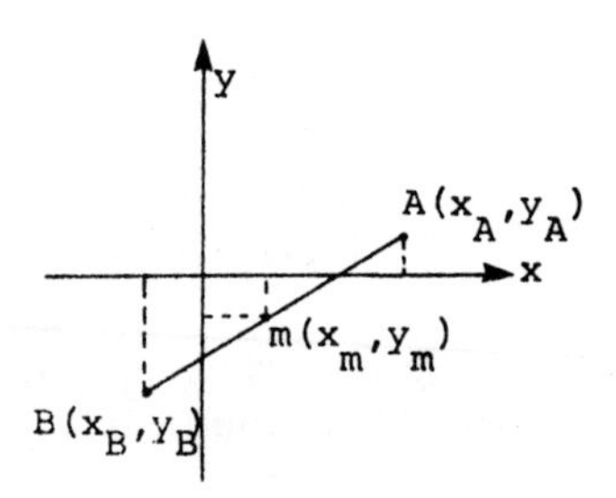

The distance between a given point $A(x_0, y_0)$ and a line ℓ, $ax + by + c = 0$, is given by the formula:

$$d = \left| \frac{ax_0 + by_0 + c}{\sqrt{a^2 + b^2}} \right|$$

11.2 SLOPE AND LINE EQUATIONS

Definition 1

The slope, m, of a non-vertical line segment determined by points $P(x_1,y_1)$ and $Q(x_2,y_2)$ is given by $m = \frac{y_2-y_1}{x_2-x_1}$.

Definition 2

The slope, m, of a non-vertical line is equal to the slope of any segment on that line.

Definition 3

The slope m of a line ℓ equals the tangent of its inclination, where the inclination α is the angle above the x-axis and is included between the line and the positive direction of the x-axis.

Definition 4

If the graph of a line crosses a coordinate axis, the point at which this occurs is called an intercept. If a line crosses the x-axis, this point is called the x-intercept. This point always has the y value, y = 0. Similarly, if a line crosses the y-axis, this point is called the y-intercept and always has the x value, x = 0.

Note: In the equation of a straight line, Y = kx + b, which will be discussed later, b is the y-coordinate of the y-intercept.

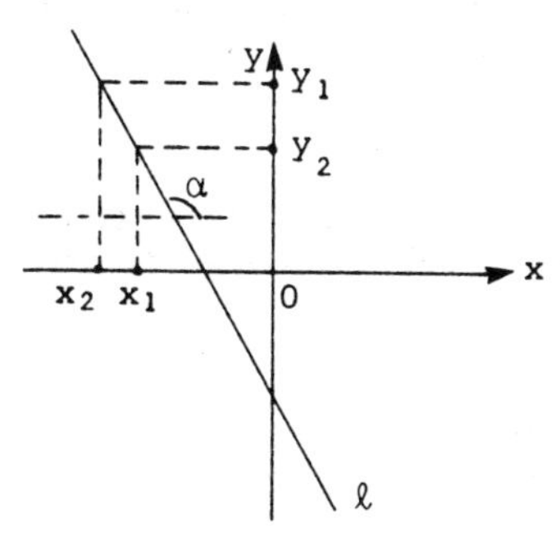

The slope of ℓ

$$m = \frac{y_2-y_1}{x_2-x_1} = \tan\alpha$$

Theorem 1

The slope of a horizontal line is zero: $m = 0$.

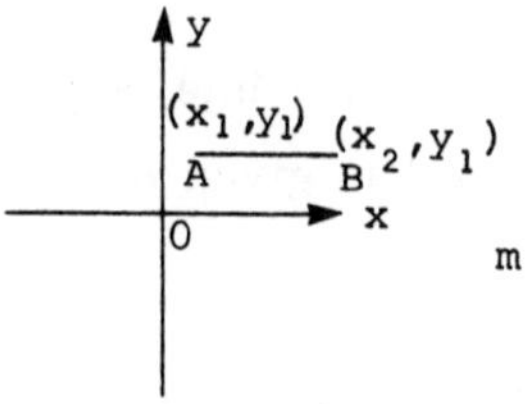

(x_1, y_1) (x_2, y_1) A B 0 x y

$$m = \frac{y_1 - y_1}{x_2 - x_1} = 0$$

Theorem 2

A vertical line has no slope.

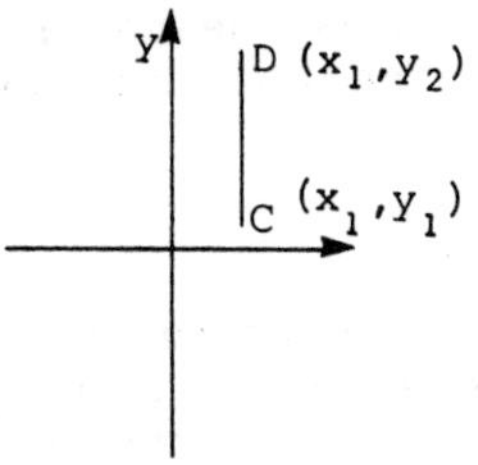

y D (x_1, y_2) C (x_1, y_1)

$$m = \frac{y_2 - y_1}{x_1 - x_1} = \frac{y_2 - y_1}{0}$$

Theorem 3

On a non-vertical line, all segments of the line have the same slope.

Theorem 4

Two non-vertical lines are perpendicular if, and only if, their slopes are negative reciprocals.

Theorem 5

Two non-vertical lines are parallel if, and only if, they have the same slope. All vertical lines are parallel.

Theorem 6

A linear equation is any equation which can be written in the form of $Ax + By + C = 0$.

Theorem 7

If the slope of a segment between a first point and second point equals the slope of the line segment between either point and third point, then the points are collinear.

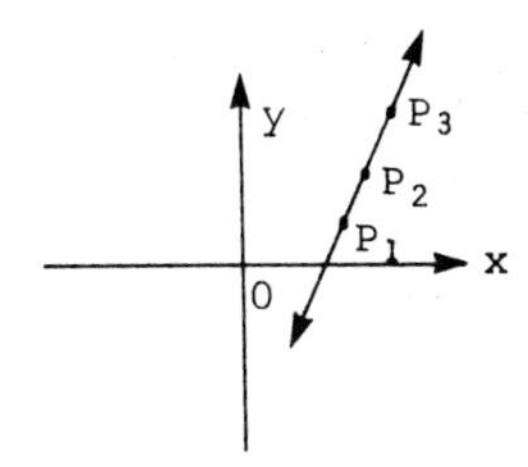

Theorem 8

The equation of the line passing through the origin and having slope k is

$$y = kx.$$

Theorem 9

The equation of the line whose y-intercept is b, and x-intercept is a, is

$$\frac{x}{a} + \frac{y}{b} = 1.$$

Theorem 10

The equation of the line with slope k and y-intercept b is

$$y = kx + b.$$

Theorem 11

The equation of the line passing through two points, $A(X_1,Y_1)$ and $B(X_2,Y_2)$, is

$$\frac{Y - Y_1}{Y_1 - Y_2} = \frac{X - X_1}{X_1 - X_2}$$

Theorem 12

The equation of the line with slope k and which passes through a point $A(X_0,Y_0)$ is

$$Y - Y_0 = k(X - X_0).$$

Theorem 13

If a line is perpendicular to the x-axis, its equation is $x = k$; k represents the x-intercept of the line.

Theorem 14

The graph of every linear equation in x and y is always a straight line.

Theorem 15

The equation of the x-axis is $y = 0$ and the equation of the y-axis is $x = 0$.

Theorem 16

The equation of a horizontal line is $y = A$ where A is the y-intercept of the line.

Theorem 17

The equation of a circle of radius r with its center at the origin is $x^2 + y^2 = r^2$.

11.3 LOCUS

Definition

A locus is the set of points, and only those points, that satisfy a given condition or a set of given conditions.

To prove a locus is correct we must prove both of the following statements:

A) If a point is on the locus, the point satisfies the given conditions.

B) If a point satisfies the given conditions, it is on the locus.

Theorem 1

The locus of points at a given distance from a fixed point, is a circle with the fixed point as its center and with radius equal to the given distance.

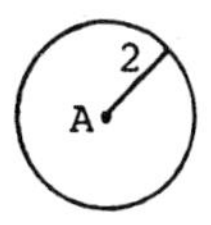

This circle represents all points that are 2 cm. from point A.

Theorem 2

The locus of points at a given distance from a given line is a pair of lines, one on each side of the given line, parallel to the given line and at the given distance from it.

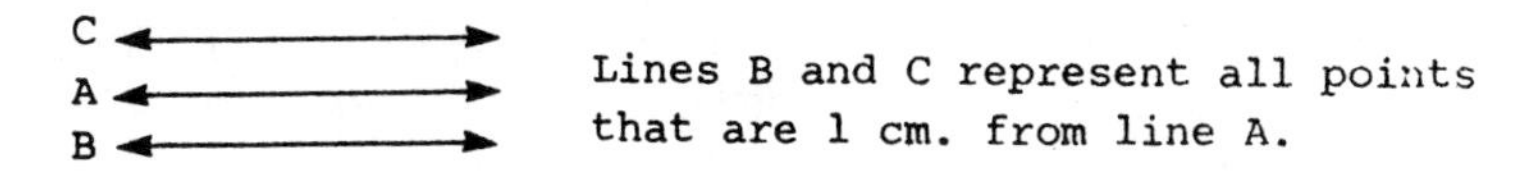

Lines B and C represent all points that are 1 cm. from line A.

Theorem 3

The locus of all points equidistant from the sides of an angle is the angle bisector.

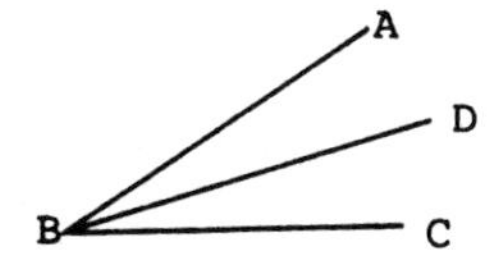

Line BD represents all points equidistant from sides AB and BC of angle ABC.

Theorem 4

The locus of all points equidistant from two points is the perpendicular bisector of the segment joining the two points.

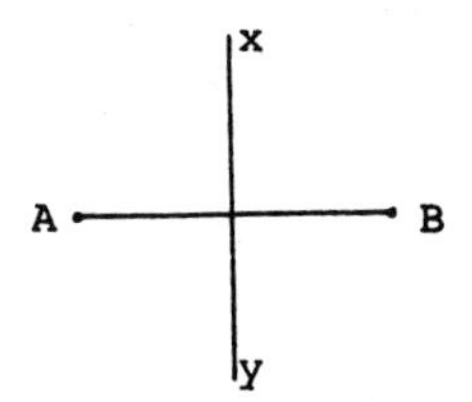

Line xy represents all points equidistant from points A and B.

Theorem 5

The locus of the vertex of the right angle of a right triangle with a fixed hypotenuse is a circle with the hypotenuse as diameter.

Theorem 6

The locus of points equidistant from two given parallel lines is a straight line parallel to both of the given lines and midway between them.

Theorem 7

The locus of points equidistant from two intersecting lines is the pair of lines which bisect the angles formed by the two intersecting lines.

Definition 1

A parabola is the locus of the points equally distant from a fixed line (called the directrix) and from a fixed point (called the focus).

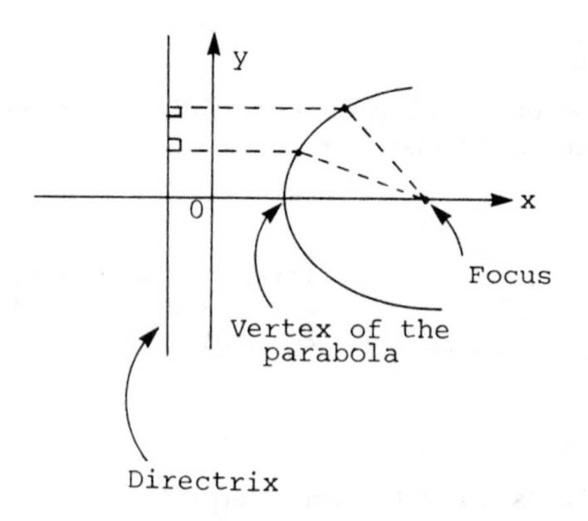

Definition 2

An ellipse is the locus of all points for which the sum of their distances from two fixed points is a constant.

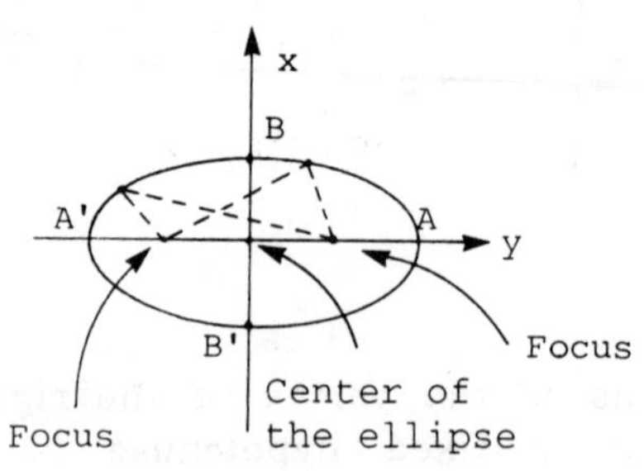

The line segment BB', which represents the perpendicular bisector of the line connecting A and A', is called the minor axis while the line segment AA' is called the major axis.

Definition 3

A hyperbola is the locus of all points for which the difference of their distances from two fixed points (called the foci) is a constant.

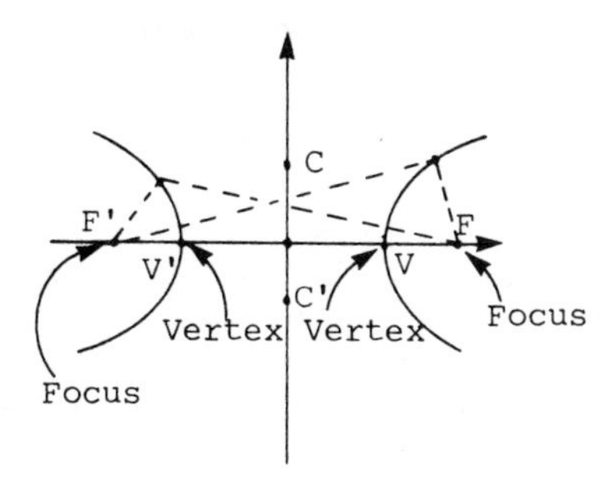

The line segment VV' is the hyperbola's transverse axis, while the segment CC' is the conjugate axis.

11.4 CONIC SECTIONS

Theorem 1

The equation of a circle is

$$(X - X_0)^2 + (Y - Y_0)^2 = r^2,$$

where (X_0, Y_0) is the coordinate of the center of the circle and r is the length of the radius.

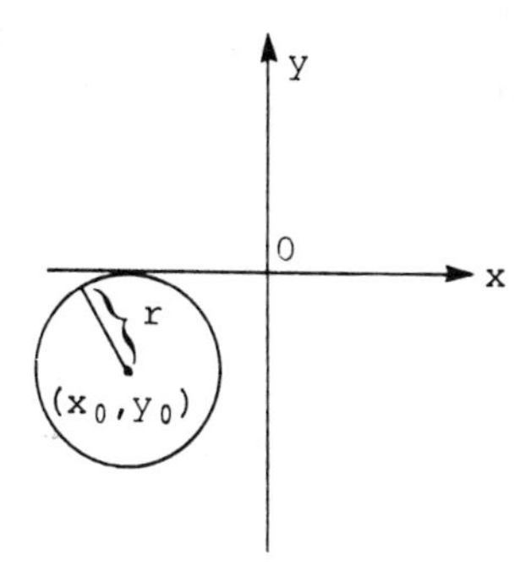

Theorem 2

A) The equation of a parabola with vertex $V(X_0,Y_0)$ and directrix d at $x = X_0 - P$ is

$$(Y - Y_0)^2 = 4P(X - X_0).$$

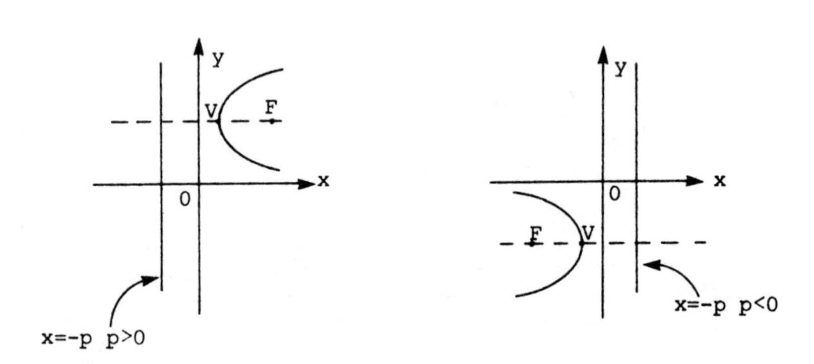

B) The equation of a parabola with vertex $V(X_0,Y_0)$and directrix d at $y = Y_0 - P$ is

$$(X - X_0)^2 = 4P(Y - Y_0).$$

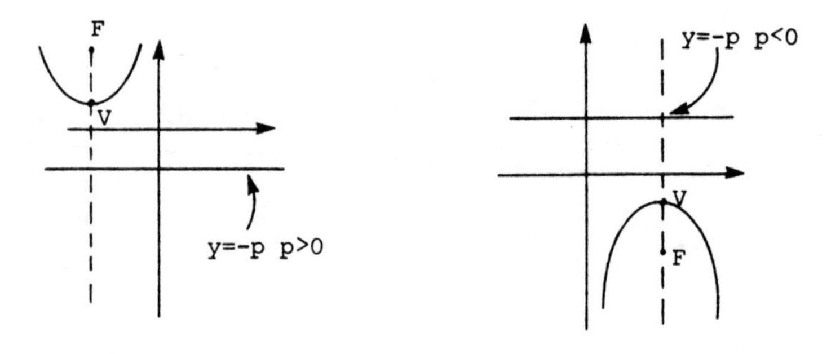

Theorem 3

The equation of an ellipse with center $C(x_0,y_0)$ and the major axis parallel to the x-axis or y-axis, respectively, is:

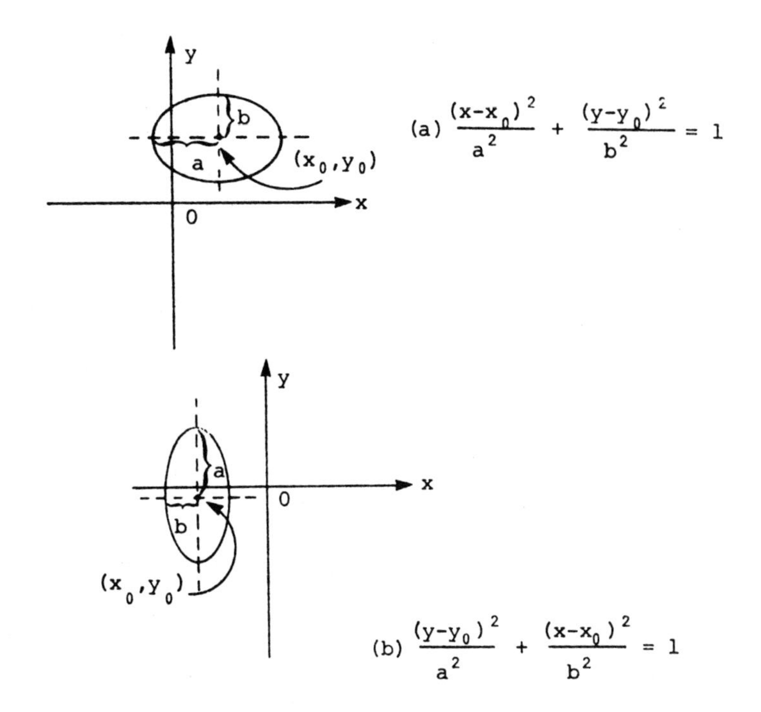

In both cases, the length of the major axis is 2a and and the length of the minor axis is 2b.

$C = \sqrt{a^2 - b^2}$ is the distance between the two foci.

Definition

$$e = \frac{c}{a} \text{ is called the eccentricity.}$$

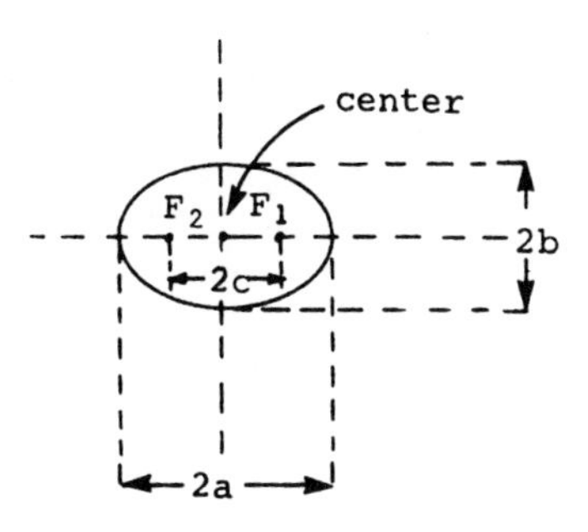

Theorem 4

The equation of a hyperbola with center $C(x_0, y_0)$

and transverse axis parallel to the x-axis or y-axis, respectively, is:

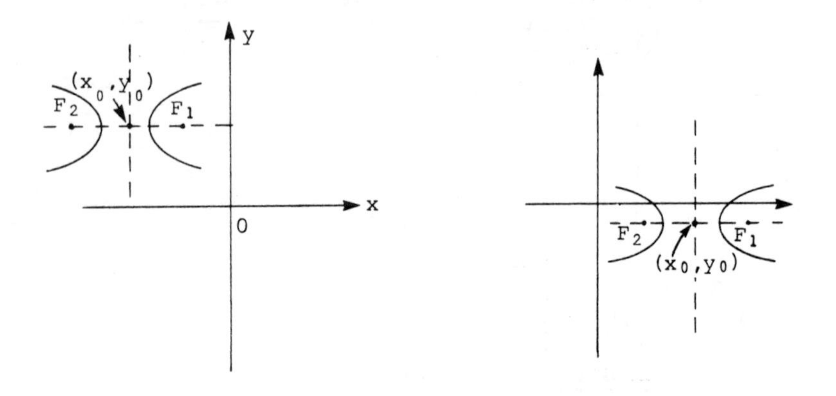

(a) $\dfrac{(x-x_0)^2}{a^2} - \dfrac{(y-y_0)^2}{b^2} = 1$ (b) $\dfrac{(y-y_0)^2}{a^2} - \dfrac{(x-x_0)^2}{b^2} = 1$

In both cases, the length of the transverse axis is 2a and the length of the conjugate axis is 2b.

$C = \sqrt{a^2 + b^2}$ is the distance between the two foci F_1 and F_2.

$$e = \frac{c}{a} \text{ is the eccentricity.}$$

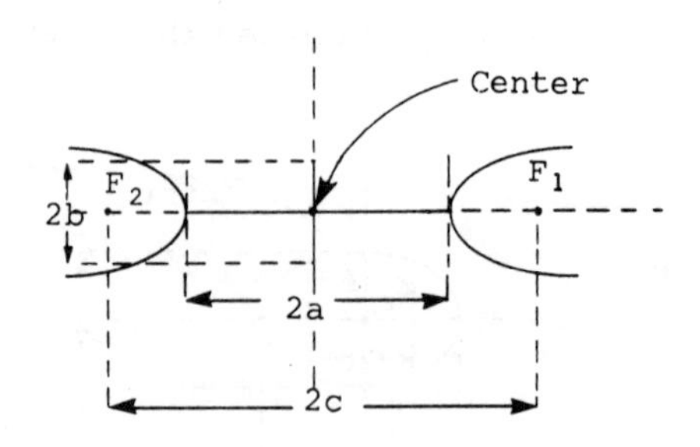

Definition 1

A cylinder is the three-dimensional figure that is generated when a rectangle is revolved about one of its sides.

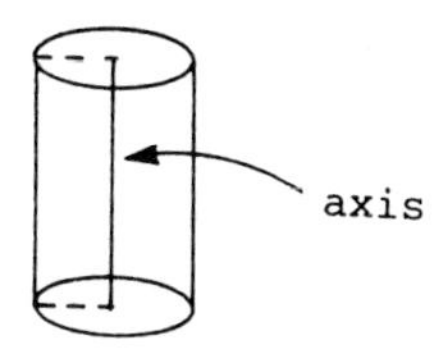

Definition 2

A cone is the three-dimensional figure that is generated when a right triangle is revolved about one of its legs.

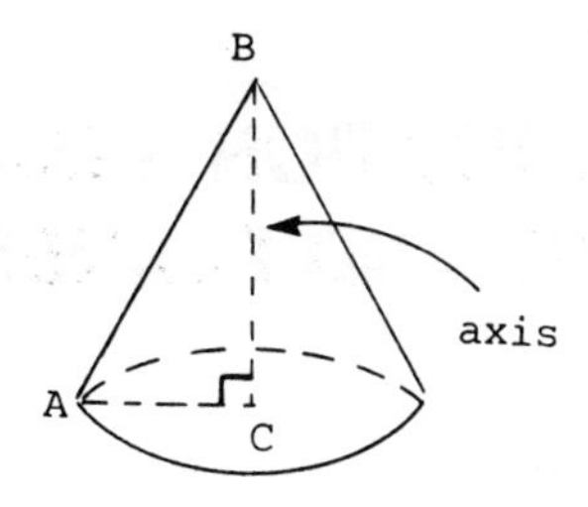

CHAPTER 12

SOLID GEOMETRY

12.1 INTRODUCTION POINTS, LINES, ANGLES AND PLANES

Definition 1

Solid geometry is the study of figures which consist of points not all in the same plane.

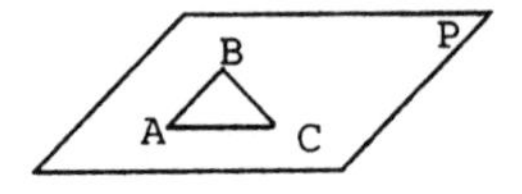

Definition 2

The intersection of a plane and a line not lying in the plane is the set of points that lie both on the line and in the plane.

Note: It will later be stated (in Postulate 2) that this set contains at most one point.

Definition 3

The intersection of two planes is the set of points that lie in both planes. Note: It will later be stated (in Postulate 3) that this set consists of points which form a straight line.

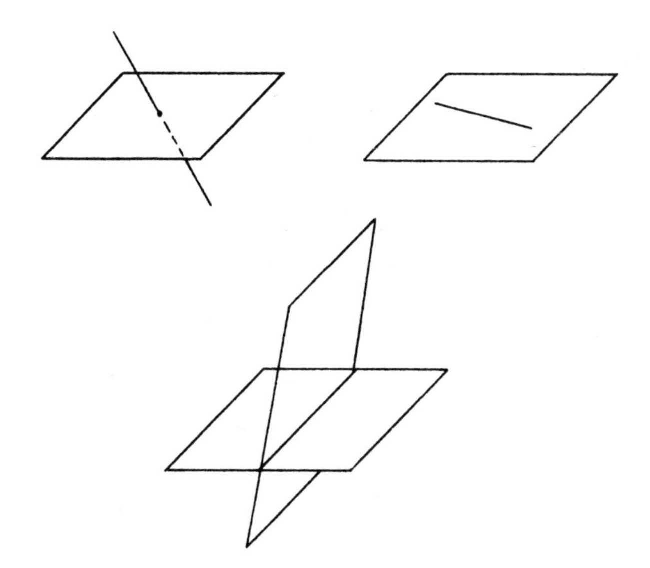

Definition 4

The intersection of a straight line and a plane is a point called the foot of the line.

Definition 5

A line is perpendicular to a given plane if it is perpendicular to every line in the plane which passes through its foot.

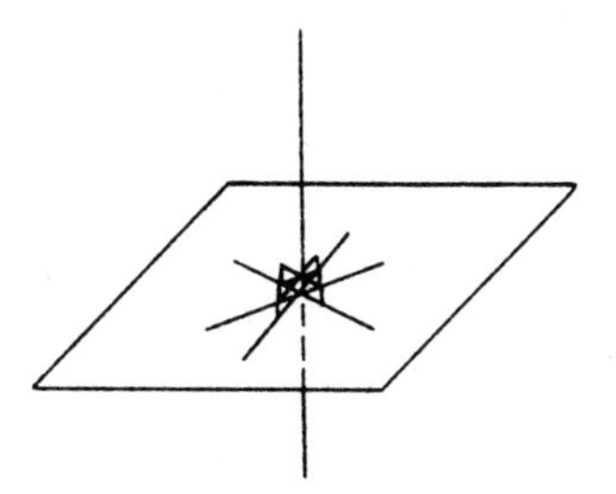

Definition 6

Lines that do not lie in the same plane are called skew lines.

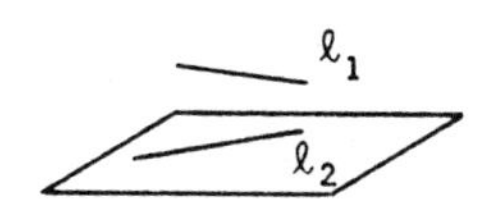

Definition 7

If a line and a plane do not intersect, then they are parallel.

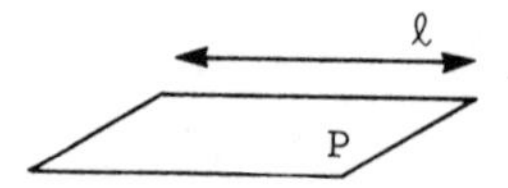

Definition 8

Two planes are parallel if they do not intersect.

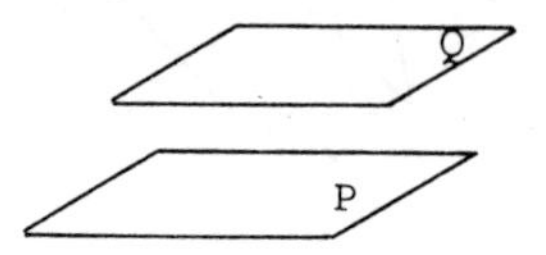

Definition 9

A straight line in a plane separates the points of the plane which are not in the line into two regions, each of which is called a half-plane.

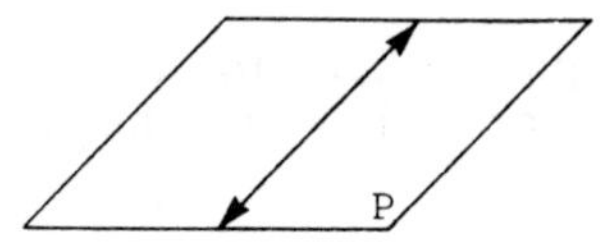

Definition 10

The set of points consisting of the union of two intersecting half-planes and their common edge is called a dihedral angle. A dihedral angle is named using 4 points, two on the common edge and two others, one unique to each half-plane.

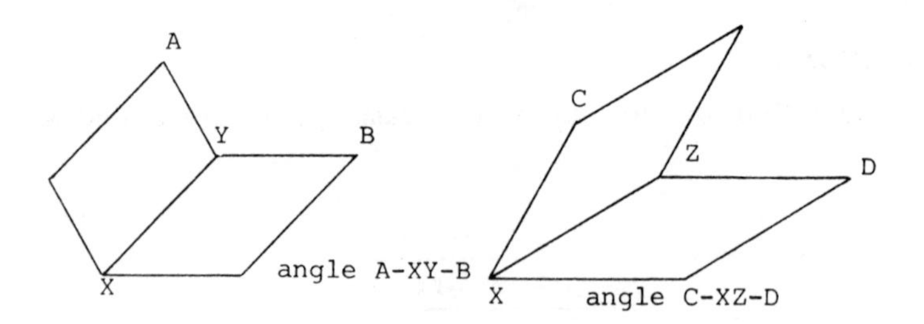

Definition 11

The angle formed by two rays, one in each face of the dihedral angle and both drawn perpendicular to the edge of the dihedral angle, is called a plane angle of the dihedral angle.

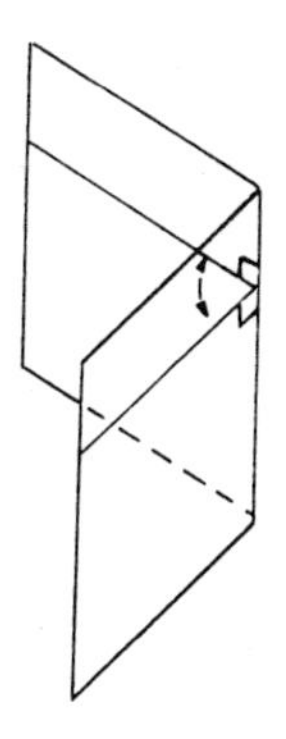

Definition 12

The measure of a dihedral angle is the measure of its plane angle.

Definition 13

Two dihedral angles with the same edge and a common face between them are adjacent dihedral angles.

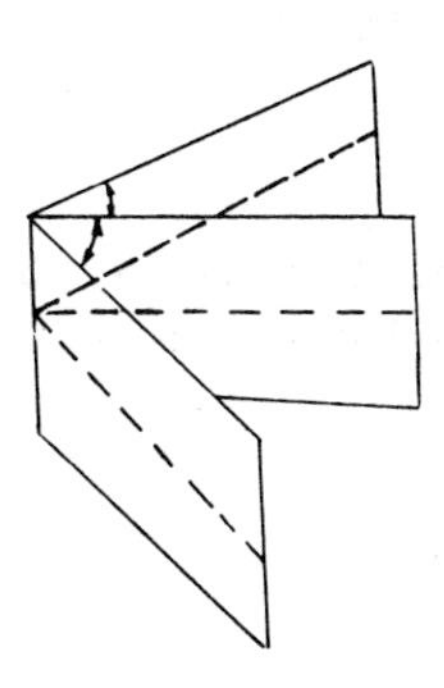

Definition 14

A dihedral angle is said to be acute, obtuse, right, or straight if its plane angle is acute, obtuse, right, or straight, respectively.

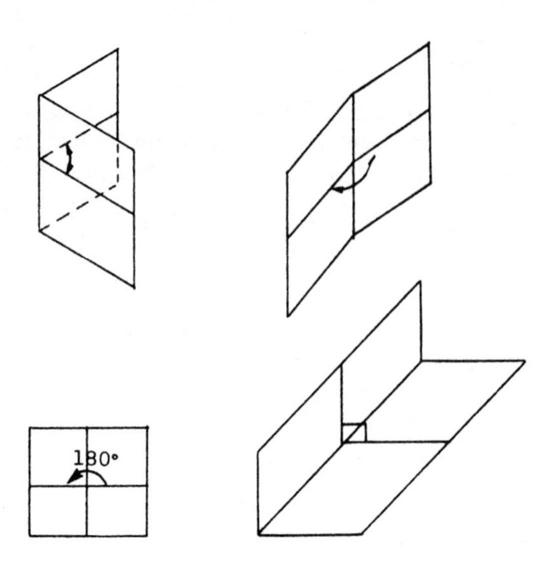

Definition 15

Two dihedral angles are complementary if their plane angles are complementary.

Definition 16

Two dihedral angles with supplementary plane angles are said to be supplementary.

Definition 17

Dihedral angles with the same measure are congruent dihedral angles.

Definition 18

Dihedral angles that have a common edge and have faces which are opposite half-planes are vertical dihedral angles.

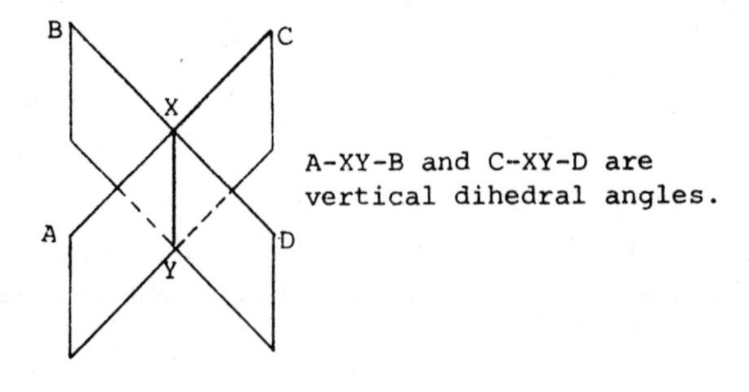

A-XY-B and C-XY-D are vertical dihedral angles.

Definition 19

If two planes intersect and form right dihedral angles, then they are perpendicular.

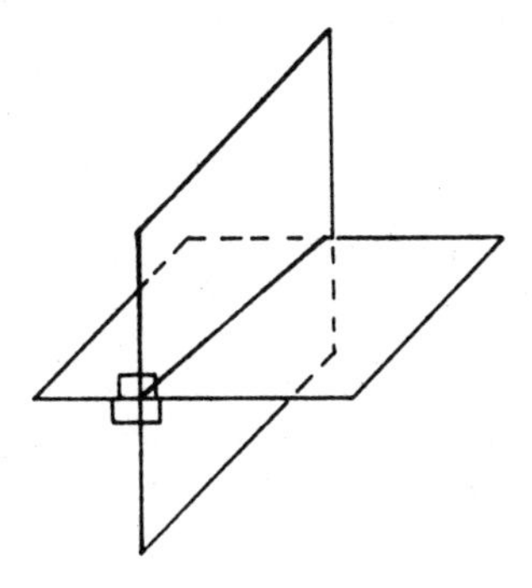

Definition 20

The distance from a point to a plane is the length of the line segment with endpoints at the given point and at the foot of the perpendicular drawn from the point to the plane.

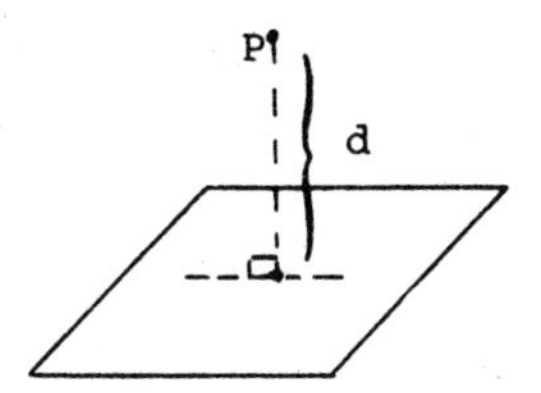

Definition 21

The distance between two parallel planes is the length of the line segment with endpoints at a given point on one plane and at the foot of the perpendicular drawn from the given point to the other plane.

Definition 22

The projection of a point onto a given plane is the foot of the perpendicular drawn from the point to the plane.

Definition 23

The projection of a line onto a plane is the set of points which are projections of all the points of the line onto the plane.

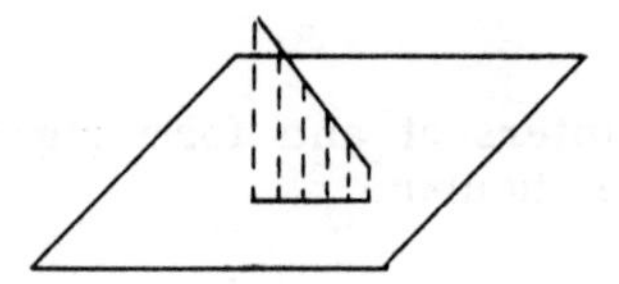

Definition 24

The angle which a line makes with a plane is the angle which it makes with its projection on the plane.

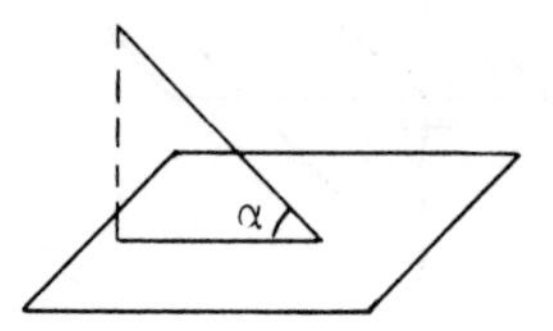

Postulate 1

A straight line lies entirely in a given plane if the line passes through two points in the plane.

Postulate 2

A plane and a line that does not lie in the plane can intersect in exactly one point.

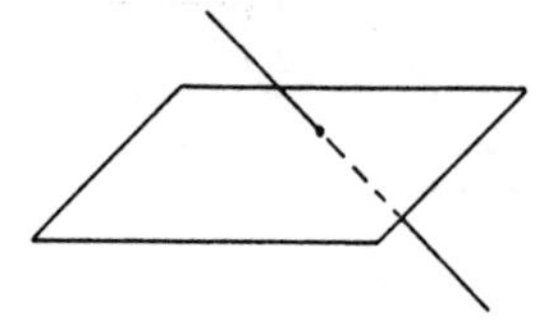

Postulate 3

The intersection of two planes can only be a straight line.

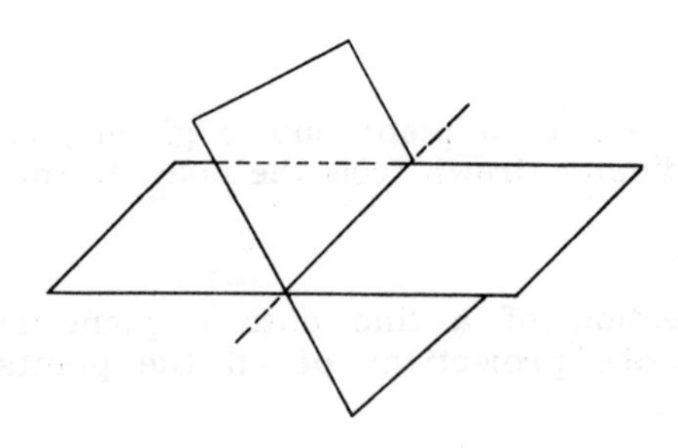

Postulate 4

Three points which are not on the same straight line determine a plane.

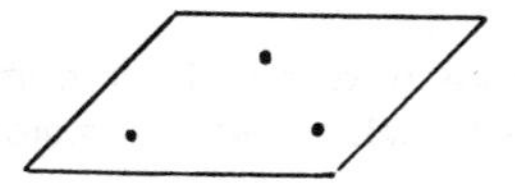

Postulate 5

If a line is perpendicular to each of two intersecting lines at their point of intersection, the line is perpendicular to the plane determined by these two intersecting lines.

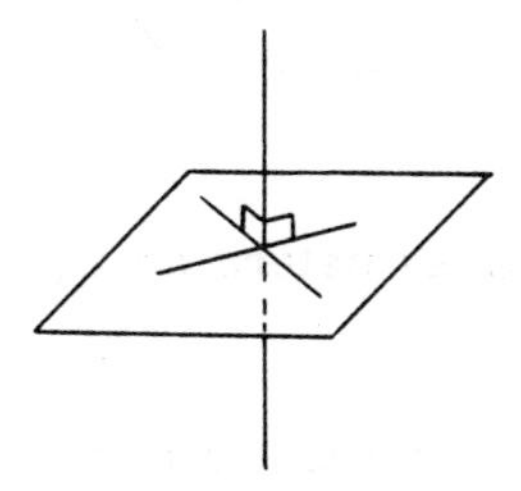

Postulate 6

At any given point in a plane there exists one and only one line perpendicular to that plane.

Postulate 7

All the perpendiculars to a line, at a point on the line, lie in a plane which is perpendicular to the line at that point.

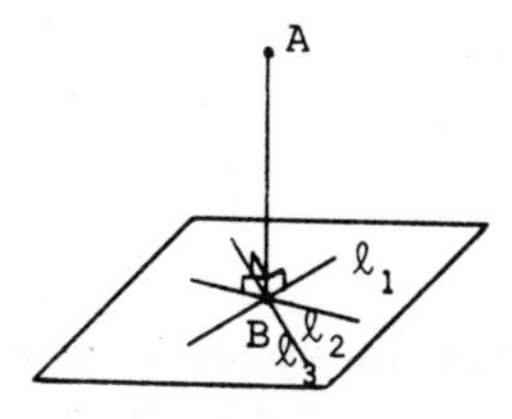

Postulate 8

For a given point outside a plane, there is one and only one line perpendicular to the plane and passing through the given point.

Postulate 9

Space contains at least four non-coplanar points.

Postulate 10

If two planes have one point in common, there exists at least two distinct points which are common to both planes.

Theorem 1

A line and a point not on that line determine a plane.

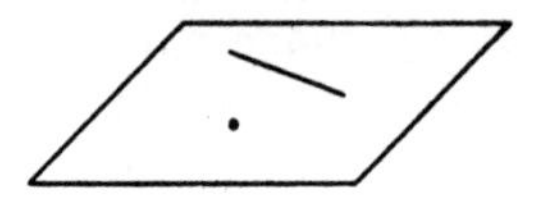

Theorem 2

Two intersecting lines determine a plane.

Theorem 3

Two parallel lines determine a plane.

Theorem 4

Two or more lines perpendicular to the same plane are parallel.

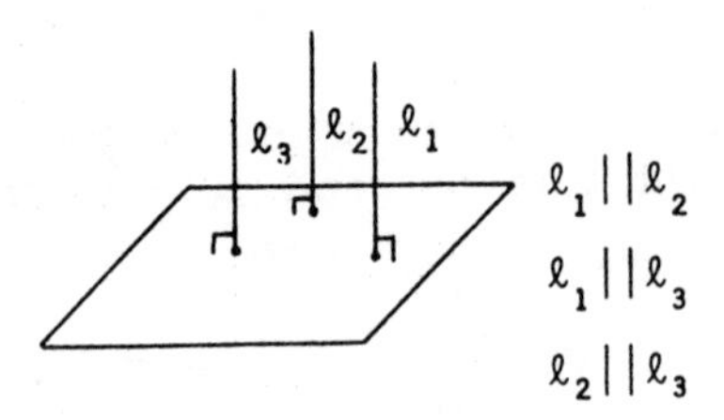

Theorem 5

If each of two lines in space is parallel to a third line, then the lines are parallel to each other.

Theorem 6

Two planes which are perpendicular to the same line are parallel.

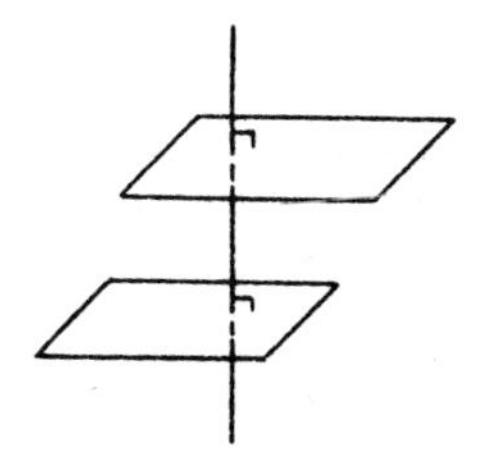

Theorem 7

Through a point outside a given plane there exists one and only one plane which is parallel to the given plane.

Theorem 8

If two parallel planes are cut by a third plane, the lines of intersection are parallel.

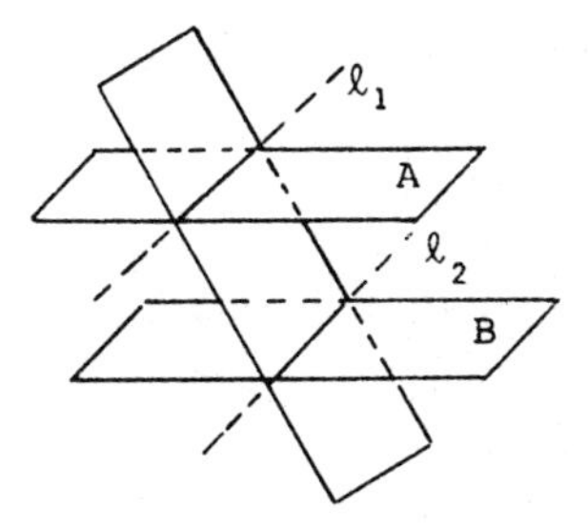

If plane A is parallel to plane B, then $\ell_1 || \ell_2$

Theorem 9

If one of two or more parallel lines is perpendicular to a given plane, the other lines are also perpendicular to the plane.

Theorem 10

If each of three non-collinear points of a plane is equidistant from two points outside the plane, then every point of the plane is equidistant from these two points.

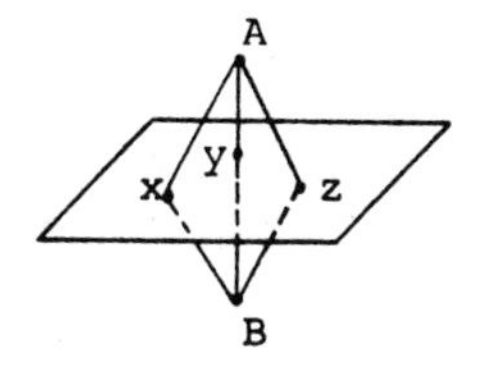

AX=BX AT=BT AZ=BZ

Therefore, the distance for every point on this plane is the same to A as it is to B.

Theorem 11

If a straight line is perpendicular to one of two or more parallel planes, it is perpendicular to all the other planes also.

Theorem 12

The distance between two parallel planes is the perpendicular distance between these two planes.

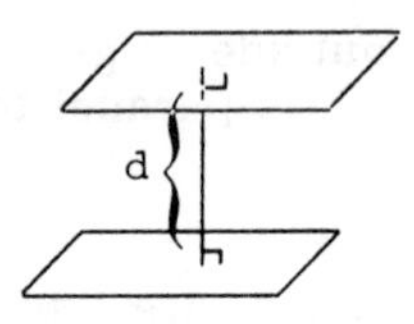

Theorem 13

Through a given straight line, an infinite number of planes may be drawn.

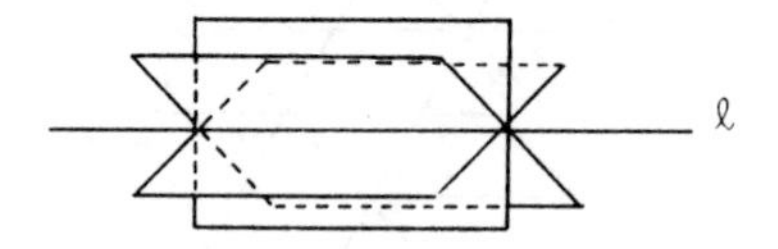

Theorem 14

If the plane angles of two dihedral angles are congruent, then the dihedral angles are congruent.

Theorem 15

If two dihedral angles are congruent, their plane angles are congruent.

Theorem 16

Two dihedral angles are congruent if they are complementary to the same dihedral angle or congruent dihedral angles.

Theorem 17

Two dihedral angles are congruent if they are supplementary to the same dihedral angle or congruent dihedral angles.

Theorem 18

The vertical dihedral angles formed when two planes intersect are congruent.

Theorem 19

Every plane which contains a line which is perpendicular to a given plane A is perpendicular to A.

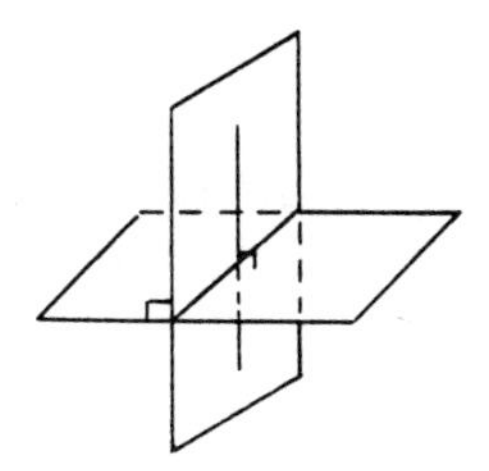

Theorem 20

Every point in a given plane that is perpendicular to a line segment at its midpoint is equidistant from the ends of the segment.

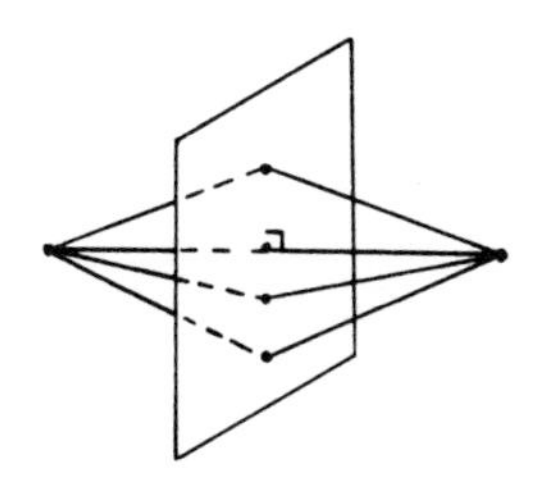

Theorem 21

All the points in a given plane which bisect a dihedral angle are equidistant from the faces of the dihedral angle.

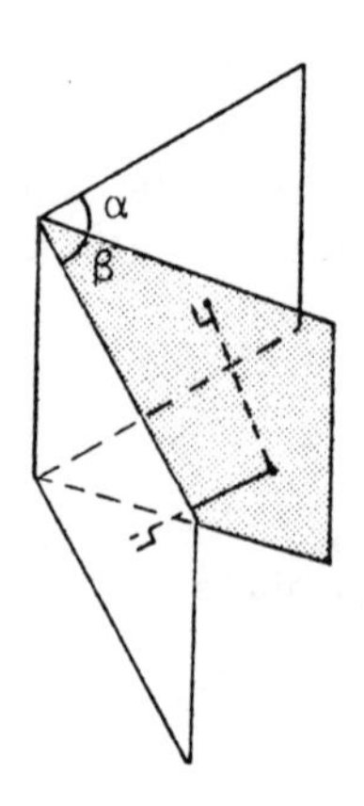

Theorem 22

A perpendicular to one of two perpendicular planes at a point of their intersection lies in the other plane.

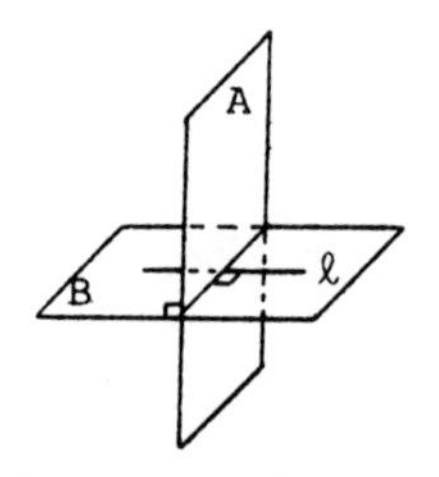

If plane A⊥ plane B and ℓ⊥ plane A, then ℓ is in plane B.

Theorem 23

The intersection of two planes which are both perpendicular to a third plane is also perpendicular to that third plane.

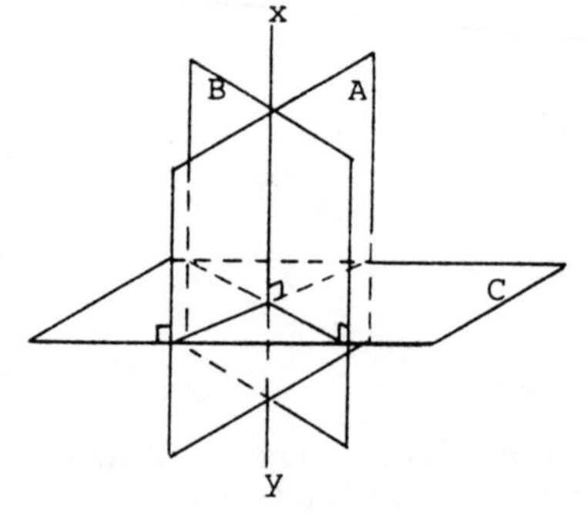

Planes A and B, which intersect at edge xy, are both perpendicular to plane C. Therefore, xy is also perpendicular to plane C.

Theorem 24

The set of points within a dihedral angle and equidistant from its faces is the plane bisecting the dihedral angle.

Theorem 25

Parallel planes are everywhere equidistant.

Theorem 26

If two planes are both parallel to a third plane, then they are parallel to each other.

12.2 POLYHEDRONS AND REGULAR POLYHEDRONS

Definition 1

A region of space enclosed by planes and curved surfaces is called a solid.

Definition 2

A polyhedral angle is the geometric figure formed by three or more planes (called faces) that intersect in the vertex of the polyhedral angle. The intersections of the faces of the polyhedral angle are called edges and the two edges of each face comprise the face angle of the polyhedral angle.

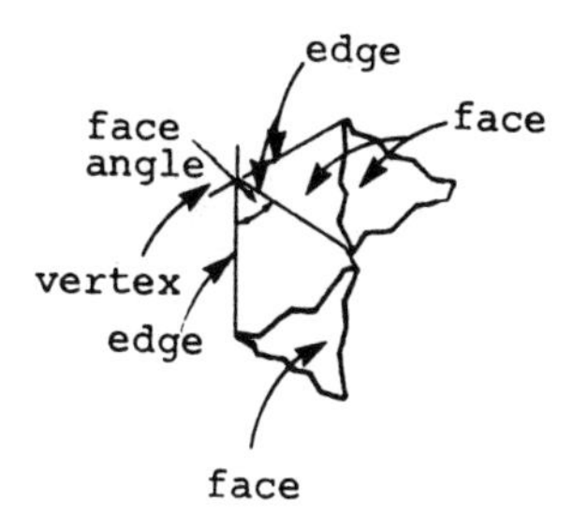

Definition 3

A polyhedral region is a solid completely bounded by portions of intersecting planes.

Definition 4

A polyhedron is the union of the bounding plane regions of a polyhedral region.

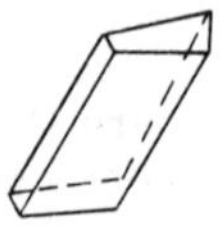

Definition 5

The line segment which joins two vertices of a polyhedron, which are not in the same face, is called a diagonal of the polyhedron.

Definition 6

In any regular polyhedron, the faces are congruent regular polygons. A regular polyhedron also has the same number of faces intersecting at each vertex.

Definition 7

There are only 5 regular polyhedrons, they are:

Tetrahedron (4 faces)

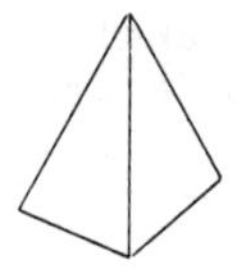

Hexahedron (6 faces) (cube)

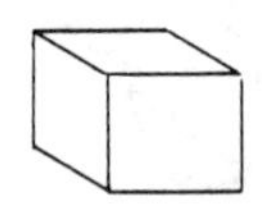

Octahedron (8 faces)

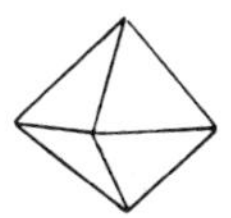

Dodecahedron (12 faces)

Icosahedron (20 faces)

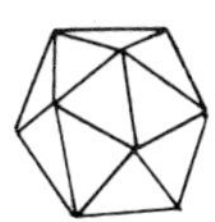

Definition 8

A prism is a polyhedron in which two faces are congruent polygons lying in parallel planes; the other faces are parallelograms.

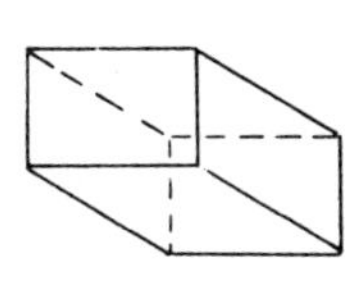

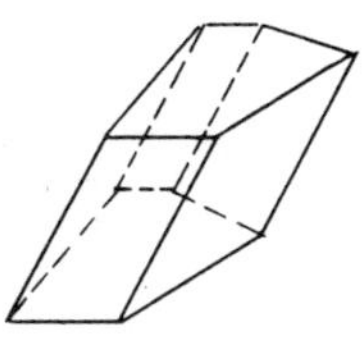

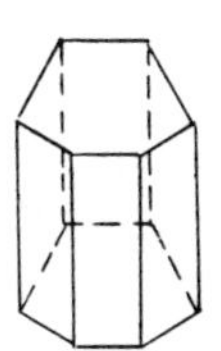

Definition 9

The perpendicular segment between the parallel planes of the bases of a prism is called the altitude of the prism.

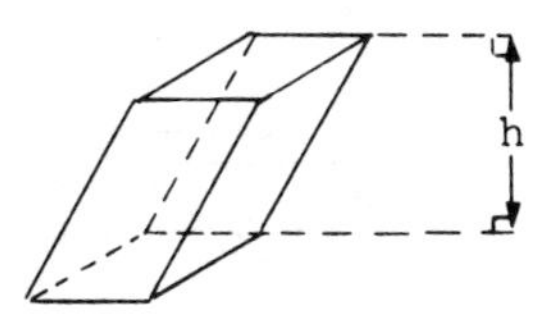

Definition 10

An oblique prism is a prism with lateral edges that are not perpendicular to the bases.

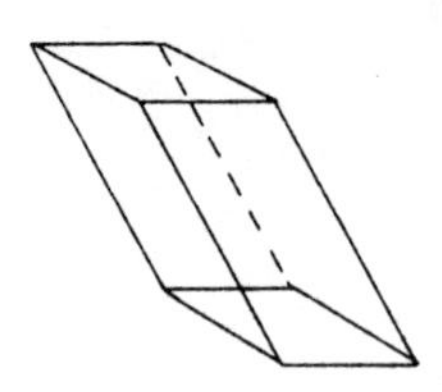

Definition 11

A right prism is a prism with lateral edges that are rectangles.

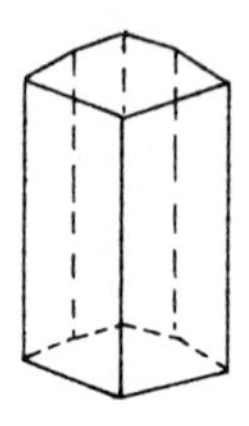

Definition 12

A regular prism is a prism with bases that are regular polygons.

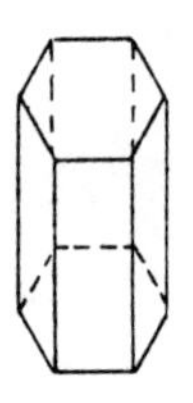

Definition 13

A prism whose bases are parallelograms is called a parallelepiped.

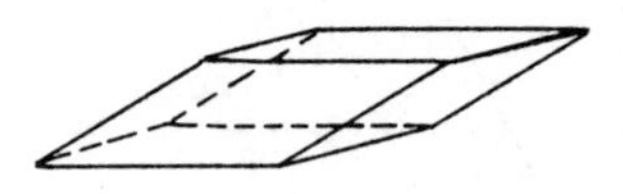

Definition 14

A parallelepiped with lateral faces and bases that are rectangles is called a rectangular parallelepiped.

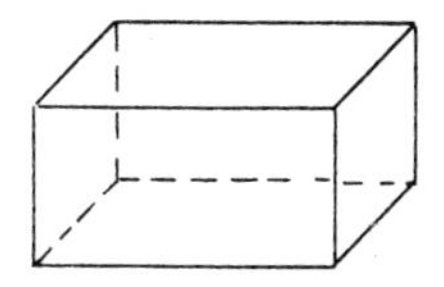

Definition 15

A rectangular parallelepiped with lateral faces and bases that are squares is called a cube.

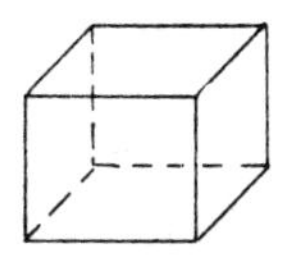

Definition 16

A polyhedron with a polygonal base and with faces that meet at a point (called the vertex) is called a pyramid.

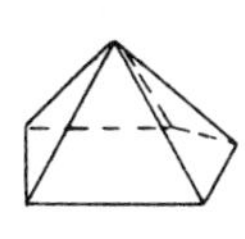

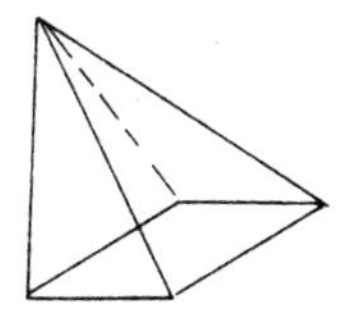

The triangular plane regions of a pyramid are called lateral faces and the intersections of the lateral faces are the lateral edges.

Definition 17

The length of the perpendicular segment from the vertex to the plane of the base of the pyramid is the altitude of the pyramid.

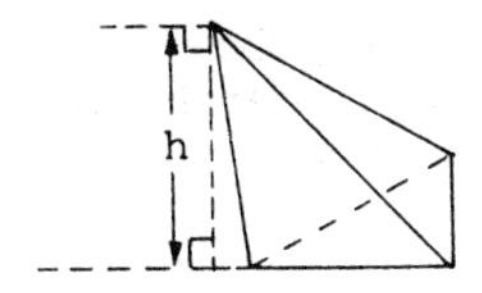

Definition 18

A pyramid with a base that is a regular polygon, and with lateral edges that are congruent is a regular pyramid.

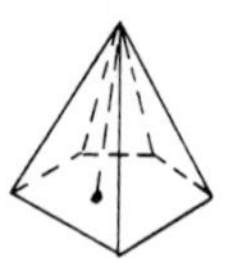
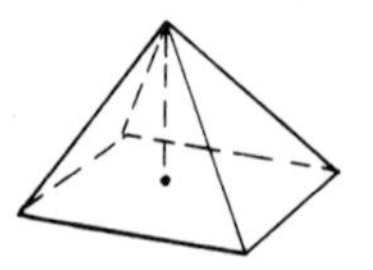

Definition 19

The slant height of a regular pyramid is the altitude drawn from the vertex to the base of any of the lateral faces.

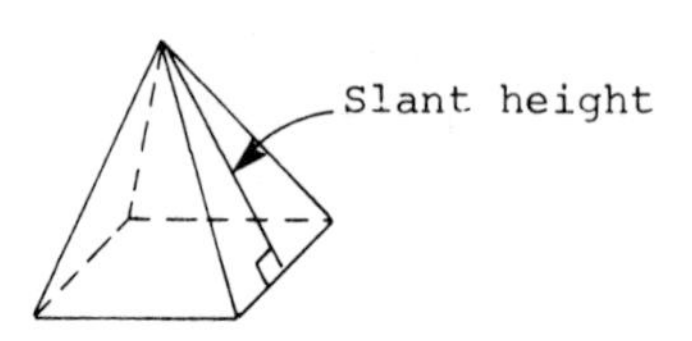

Definition 20

The part of a pyramid which remains after a plane has cut off the top of the pyramid parallel to the base, is called a frustum of a pyramid.

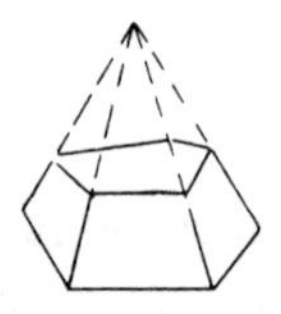

Definition 21

The section of a polyhedron is the polygonal region formed by the intersection of a polyhedron and a plane passing through it.

Definition 22

A right section of a prism is the section formed by a plane which cuts all the lateral edges of the prism and is perpendicular to one of them.

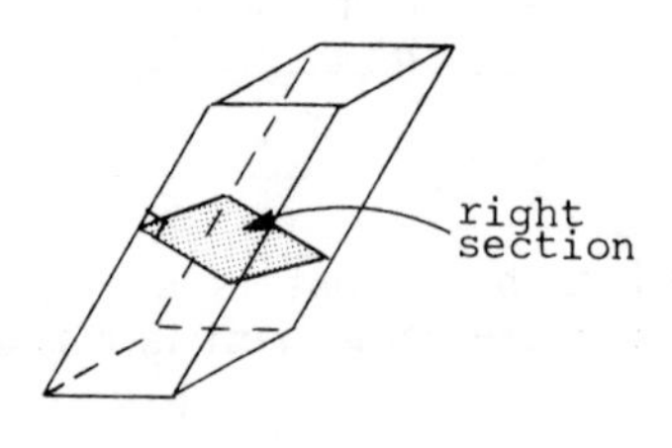

Theorem 1

The sum of the measures of any two face angles of a trihedral angle is greater than the measure of the third face angle.

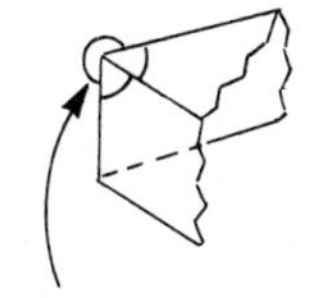

Theorem 2

The sum of the measure of the face angles of any convex polyhedral angle is less than 360°.

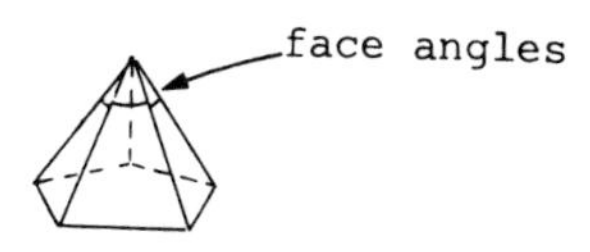

Theorem 3

The lateral edges of a prism are congruent and parallel.

Theorem 4

Every section of a prism made by a plane parallel to the bases is congruent to the bases.

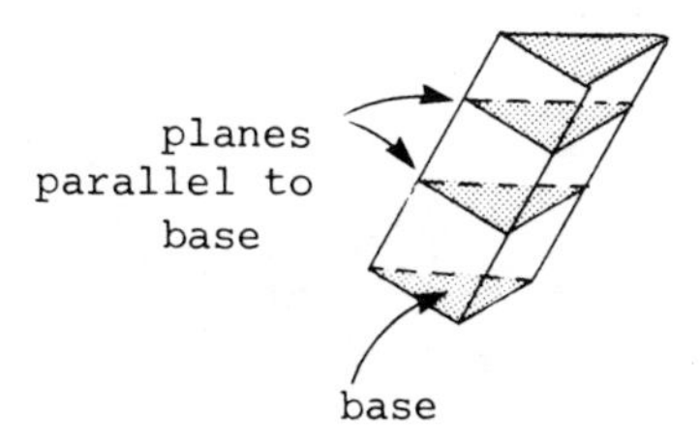

Theorem 5

The lateral edges of a regular pyramid are congruent.

Theorem 6

The lateral edges of a regular pyramid form congruent isosceles triangles.

Theorem 7

If a pyramid is cut by a plane parallel to its base, the cross section is a polygon similar to the base, and the lateral edges and altitude are divided proportionally.

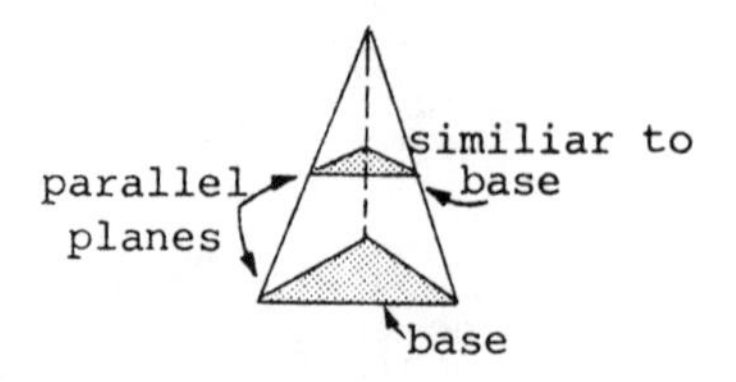

12.3 CYLINDERS, CONES AND SPHERES

Definition 1

A sphere is the set of points in space at a given distance from a given point, called the center of the sphere.

Definition 2

The distance from a given point to a sphere is the length of the line segment with endpoints that are the given point and the point of intersection of the sphere with the line that passes through the center of the sphere and the given point.

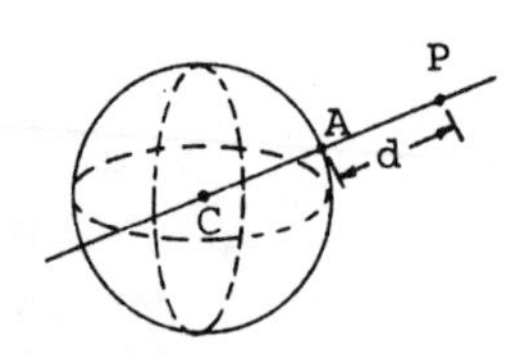

Definition 3

A great circle of a sphere is any circle lying on the sphere and having the same radius as the sphere.

Any circle on the sphere with radius less than the radius of the sphere is called a small circle of the sphere.

Definition 4

A radius of a sphere is a line segment joining the center and any point of the sphere.

A diameter of a sphere is a segment passing through the center and having its endpoints on the sphere.

The diameter of a sphere is twice as long as the radius of the sphere.

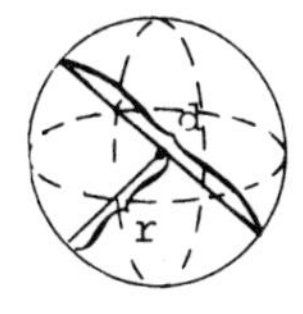

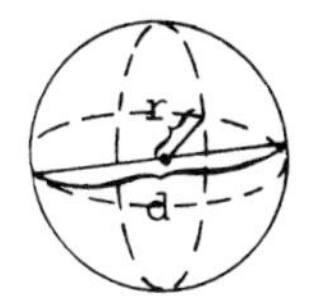

Definition 5

Congruent spheres are spheres with congruent radii.

Definition 6

A tangent to the sphere is any line which is perpendicular to a radius of a sphere at the outer extremity of the radius.

Definition 7

A cylindrical surface is the set of all lines (each is called an element of the cylindrical surface) parallel to a given line and intersecting a given curve (called the directrix) in a plane that does not contain the given line.

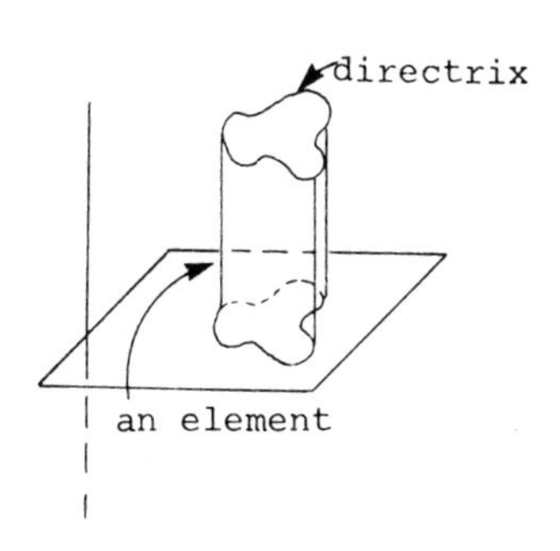

Definition 8

The portion of a closed cylindrical surface between two

parallel planes, together with the portions of the planes enclosed by the surface, is called a cylinder.

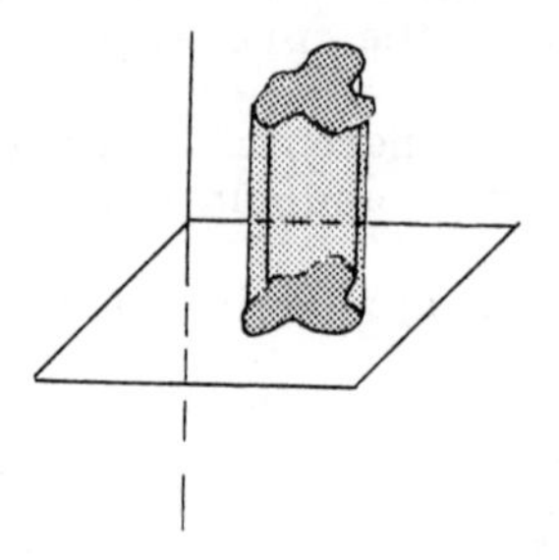

Definition 9

A solid with bases that are parallel circles and with cross sections parallel to the bases that are also circles is called a circular cylinder (or a cylinder).

Definition 10

The altitude of a cylinder is the distance between two bases of the cylinder.

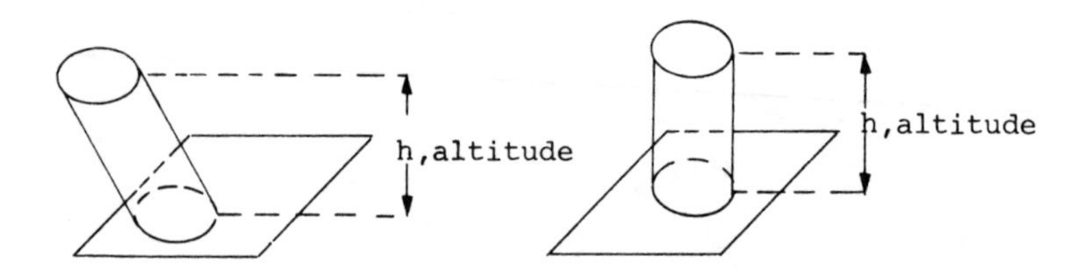

Definition 11

A cylinder of revolution, or a right circular cylinder, is formed by revolving a rectangle about one of its two dimensions as an axis.

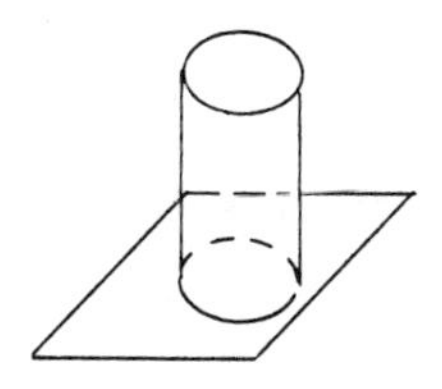

Definition 12

An oblique circular cylinder is a cylinder whose bases are congruent circles which lie in parallel planes.

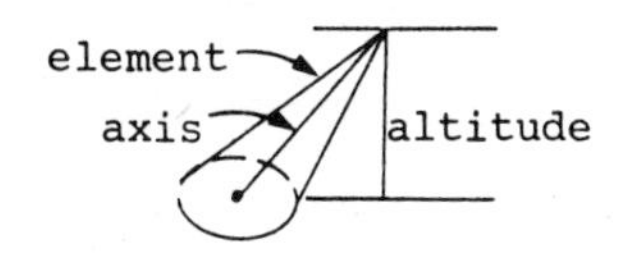

Definition 13

The line segment joining the centers of the bases of a cylinder is called the axis of the cylinder.

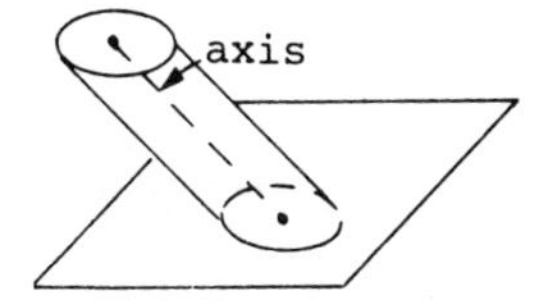

Definition 14

A conical surface is the set of all lines intersecting a given plane curve called the directrix and passing through a fixed point called the vertex which is not in the plane of the directrix.

The parts of a conical surface on each side of the vertex are called nappes of the conical surface.

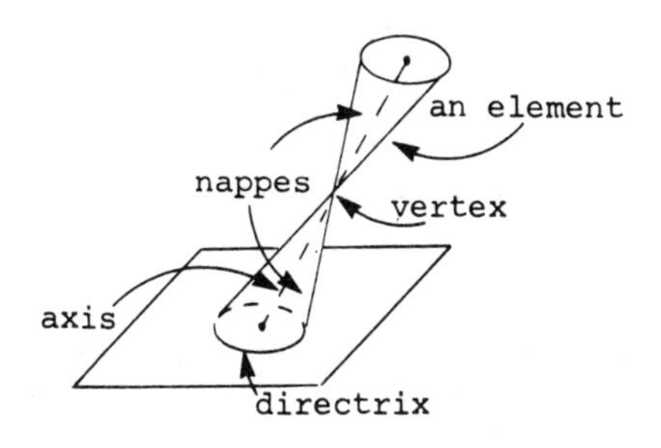

Definition 15

A cone is the part of a closed conical surface between the vertex and a plane intersecting one nappe of the conical surface.

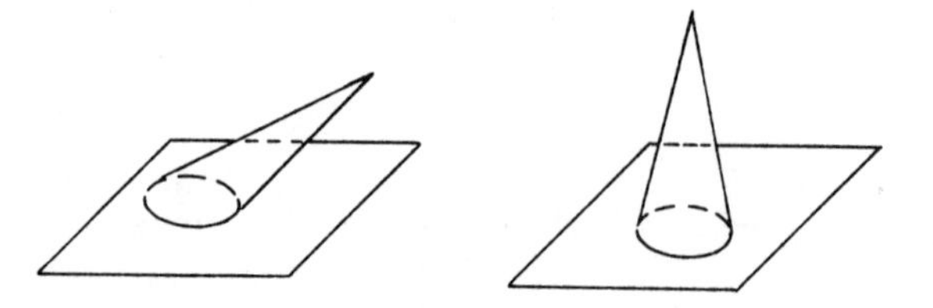

Definition 16

A circular cone whose axis is perpendicular to the plane of the base is called a right circular cone, or a cone of revolution, since it can be formed by revolving a right triangle about one of its legs.

The length of any element of the right circular cone is its slant height.

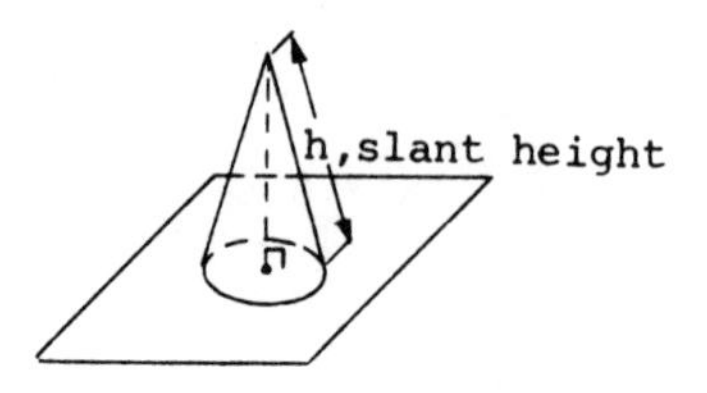

Definition 17

An oblique circular cone is a circular cone with an axis that is not perpendicular to the plane of the base of the cone.

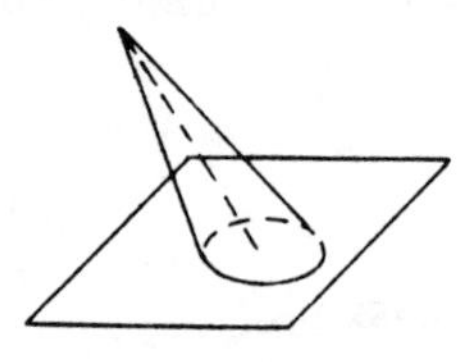

Definition 18

A frustum of a cone is the part of a cone which remains after the top of the cone has been cut off by a plane parallel to the base of the cone.

Any plane perpendicular to the axis of the cone cuts a section that is a circle. Incline the plane a bit, and the section formed is an ellipse. Tilt the plane still more until it is parallel to a ruling of the cone, and the section is a parabola. Let the plane cut both nappes, and the section is a hyperbola, a curve with two branches. It is apparent that closed orbits are circles or ellipses. Open or escape orbits are parabolas or hyperbolas.

The conic sections can be classified by means of their eccentricity. If we represent the eccentricity by e, then a conic section is

a circle if $e = 0$

an ellipse if $0 < e < 1$

a parabola if $e = 1$

a hyperbola if $e > 1$.

In actual practice, orbits that are exactly circular or parabolic do not exist because the eccentricity is never exactly equal to 0 or 1.

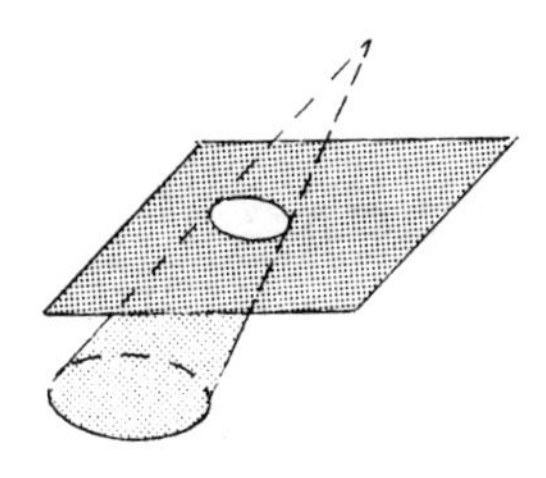

Theorem 1

The intersection of the elements of a right circular cylinder and a plane parallel to the bases of the cylinder is a circle.

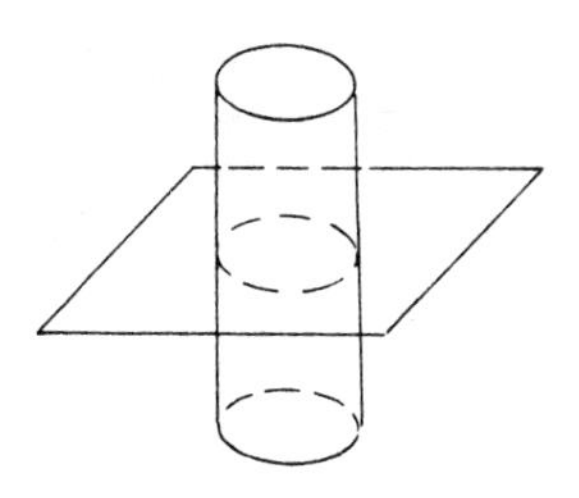

Theorem 2

The intersection of the elements of a right circular cone and a plane which is parallel to the base, but does not contain the vertex of the cone, is a circle.

Theorem 3

If a plane intersects a sphere in more than one point the intersection is a circle.

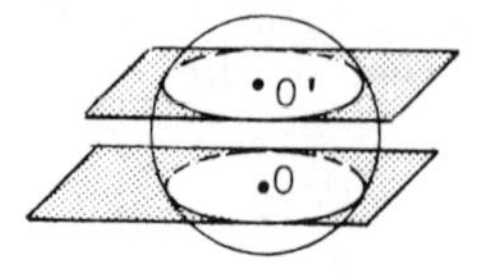

Theorem 4

The bases of a given cylinder are congruent.

Theorem 5

Every section of a cylinder made by a plane parallel to the bases is congruent to the bases of the cylinder.

12.4 SURFACE AREAS

Definition

The total area of a polyhedron is the sum of the areas of all its faces.

Theorem

The area of the surface generated by a line segment revolving about an axis in its plane, but not perpendicular to the axis nor crossing it, is equal to the product of the projection of the line segment onto the axis and the circumference of the circle with radius that is the perpendicular to the segment drawn from its midpoint to the axis.

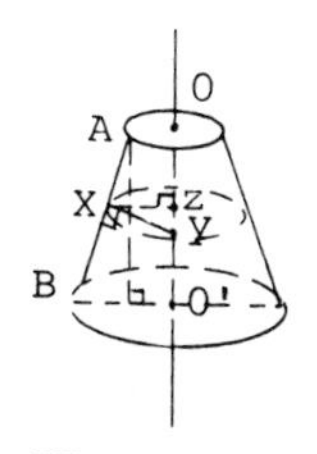

AX=XB
Xy⊥AB
S= OO'×2π×Xy

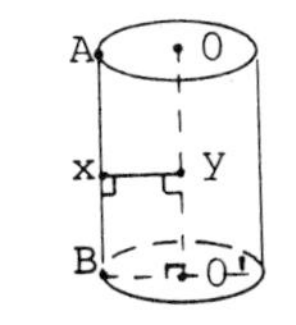

Ax=xB , xy⊥AB
S= OO'×2π×xy

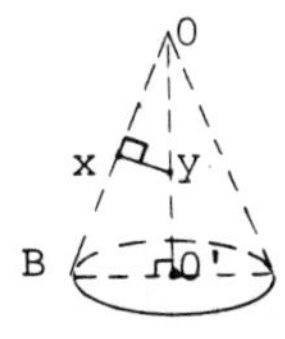

Ox=xB , xy⊥BO
s= OO'×2π×xy

I. Prisms

Definition 1

The lateral area of a prism is the sum of the areas of its lateral faces.

Definition 2

The total area of a prism is the sum of its lateral area and the areas of its two bases.

Theorem 1

The lateral area of a prism is equal to the product of the perimeter of a right section and the length of a lateral edge.

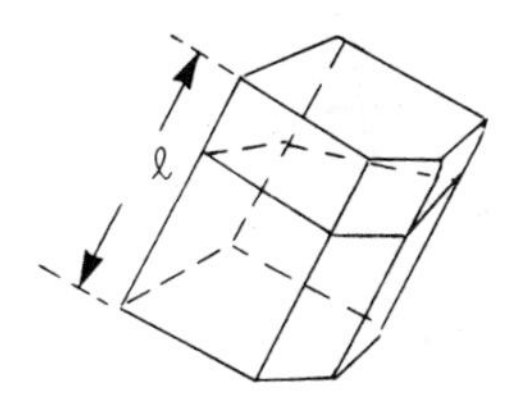

If the perimeter of the prism is P, then its lateral area is A= Pℓ.

Theorem 2

The lateral area of a right prism is equal to the product of the perimeter of one of its bases and its altitude.

Theorem 3

All cross sections of a prism have equal areas.

II. Pyramids

Definition 1

The lateral area of a pyramid is the sum of the areas of its lateral faces.

Definition 2

The total area of a pyramid is the sum of its lateral area and the area of its base.

Theorem 1

The lateral area of a regular pyramid is equal to one-half the product of its slant height and the perimeter of its base.

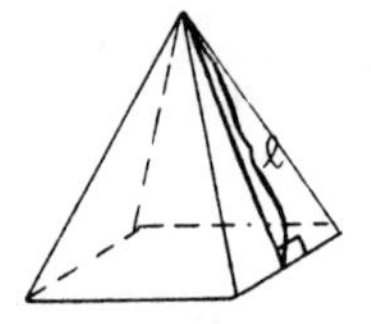

If the perimeter of the pyramid base is b, then its lateral area is $A = \frac{1}{2}b\ell$.

Theorem 2

The corresponding sections parallel to the bases of two pyramids at equal distances from the vertices of the pyramids have equal areas if the pyramids have congruent altitude, and bases with equal areas.

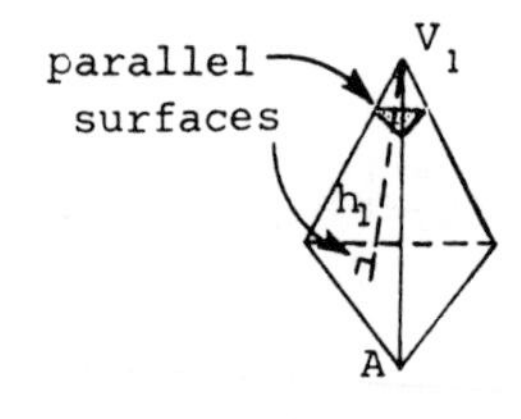

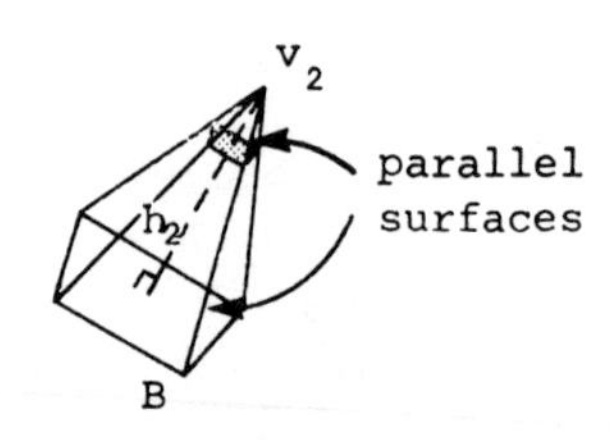

Pyramids A and B have equal heights h_1 and h_2, and their bases have equal surface areas. Therefore, any cross section of A parallel to its base will be equal in area to the cross section of B, that is both parallel to its base and also the same distance form its vertex as the cross section of A is from A's vertex.

III. Cylinders

Definition 1

The lateral area of a cylinder is equal to the area of its curved cylindrical surface.

Definition 2

The total area of a cylinder is equal to the sum of its lateral area and the areas of its bases.

Theorem

The lateral area of a right circular cylinder is equal to the product of the circumference of its base and the length of its altitude.

Postulate

The lateral area of a circular cylinder is equal to the product of an element and the perimeter of a right section.

IV. Cones

Definition 1

The lateral area of a cone is equal to the area of its curved conical surface.

Definition 2

The total area of a cone is equal to the sum of its lateral area and the area of its base.

Postulate

The lateral area of a right circular cone is equal to one-half the product of the circumference of its base and the length of its slant height.

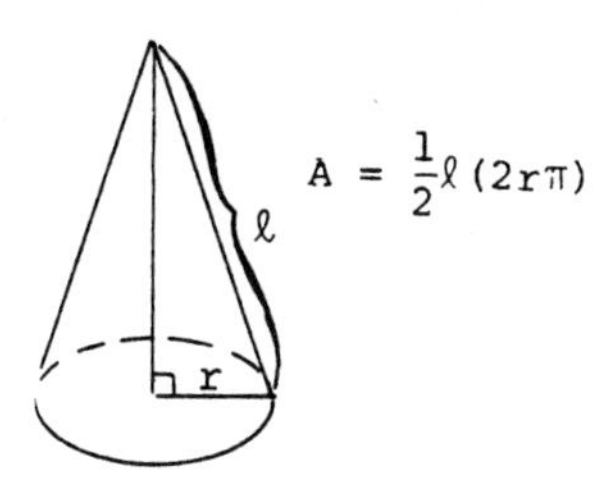

Theorem

The lateral area of a frustum of a right circular cone is

one-half the product of the slant height and the sum of the circumferences of the two bases.

$$A = \frac{1}{2}\ell[2\pi r_1 + 2\pi r_2]$$

V. Spheres

Definition

The area of a sphere is equal to the area of its curved spherical surface.

Postulate

The area of a sphere is equal to four times the area of one of its great circles.

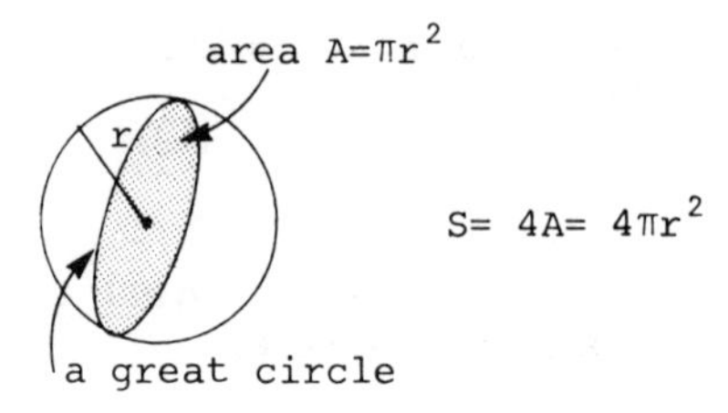

12.5 VOLUMES

Definition

The volume of a solid is the number of units of volume contained in the interior of the solid.

Postulate 1

If two solid regions have the same altitude, and if sections made by planes parallel to the base of each solid and at the same distance from each base are always equal in area, then the volumes of the solid regions are equal.

Postulate 2

The volume of a right prism is equal to the product of the area of its base and the measure of its altitude.

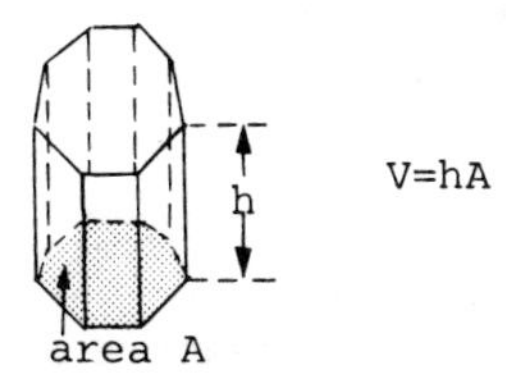

Postulate 3

The volume of any pyramid is equal to one-third the product of the area of its base and the measure of its altitude.

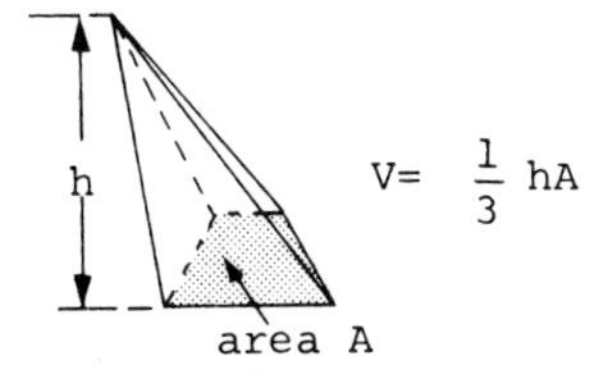

Postulate 4

The volume of a right circular cylinder is equal to the product of the area of its base and the length of its altitude.

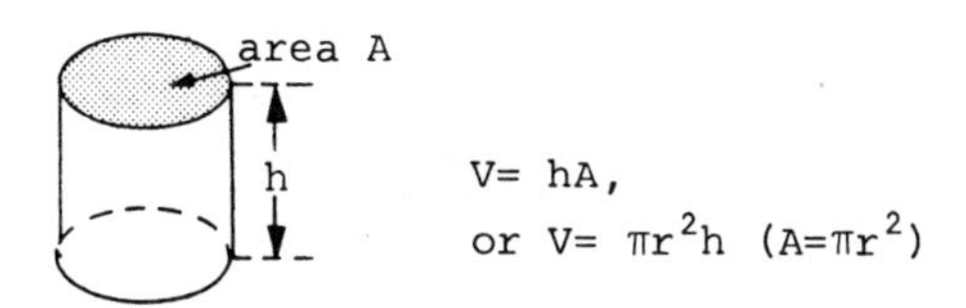

Postulate 5

The volume of a right circular cone is equal to one-third the product of the area of its base and the length of its altitude.

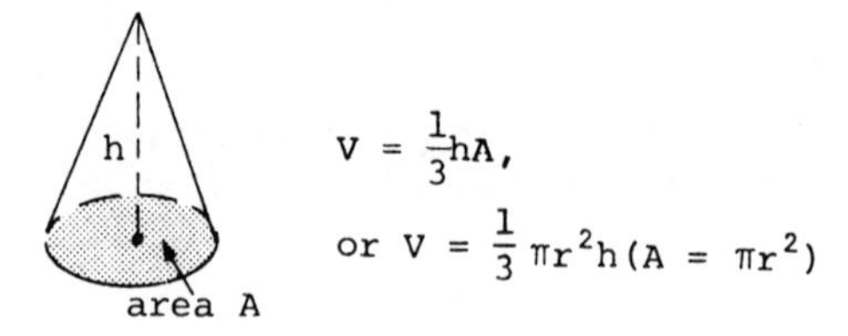

Postulate 6

The volume of a sphere is equal to one-third the product of its area and the measure of its radius.

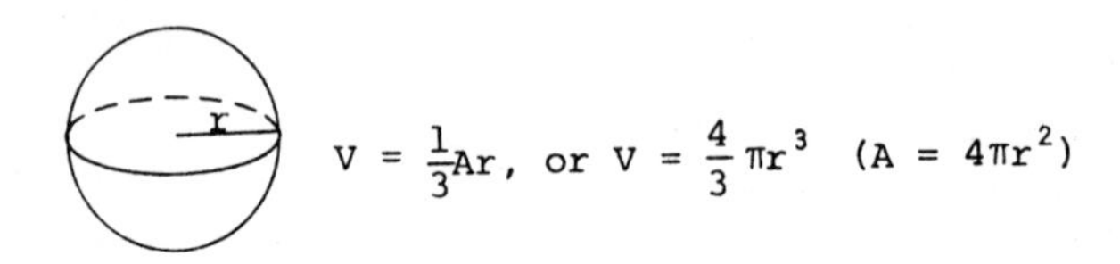

Theorem 1

The volume of a cube equals the cube of the measure of an edge.

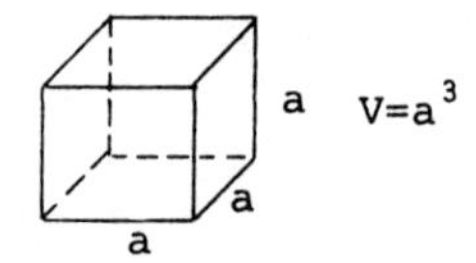

Theorem 2

The volume of a rectangular solid is equal to the product of its length, width and height.

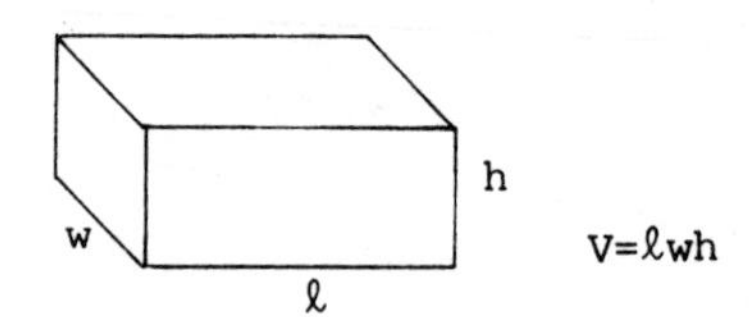

Theorem 3

Two prisms have equal volumes if their bases have equal areas and their altitudes have equal measures.

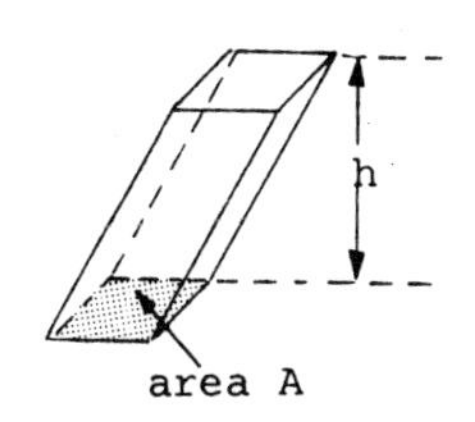

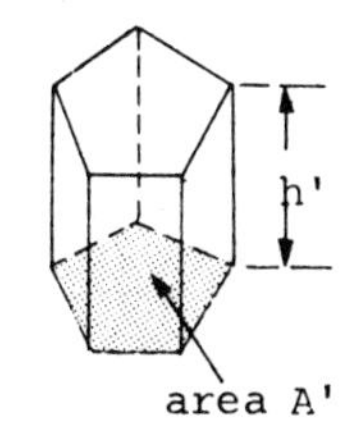

If A=A' and h=h' then V=V'

Theorem 4

The volume of any prism is the product of the area of its base and its altitude.

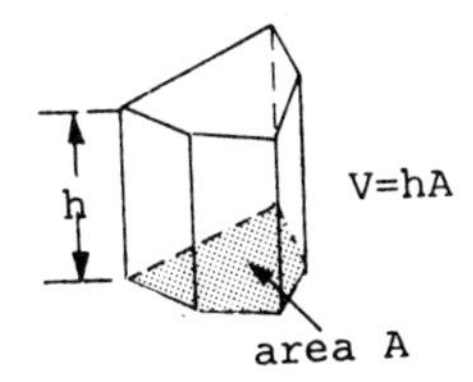

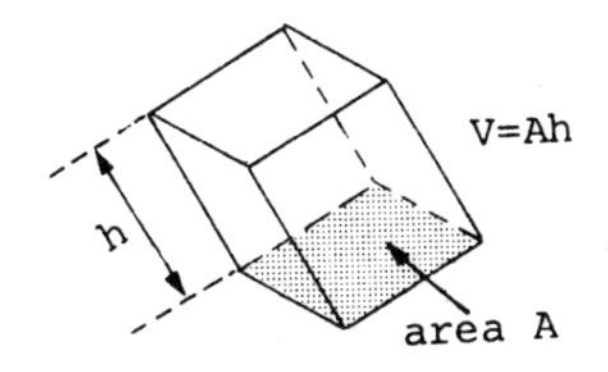

Theorem 5

If two pyramids have congruent altitudes and bases with equal areas, then they have equal volumes.

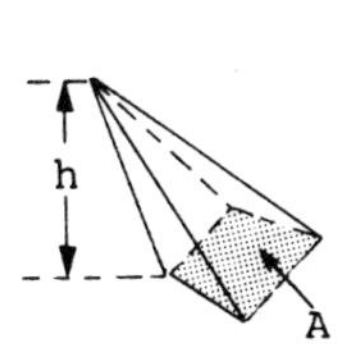

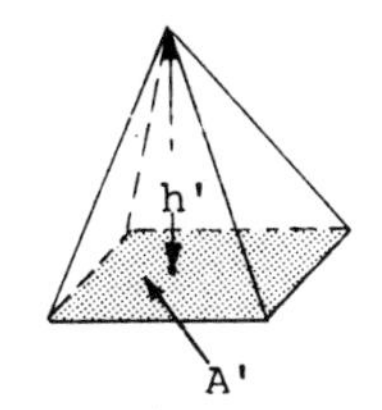

If A=A' , h'=h then V=V'

Theorem 6

The volume of any cylinder is the product of the area of its base and the altitude of its cylinder.

$$V = Ah$$

Theorem 7

The volume of any circular cone is equal to one-third the product of the area of its base and the measure of its altitude.

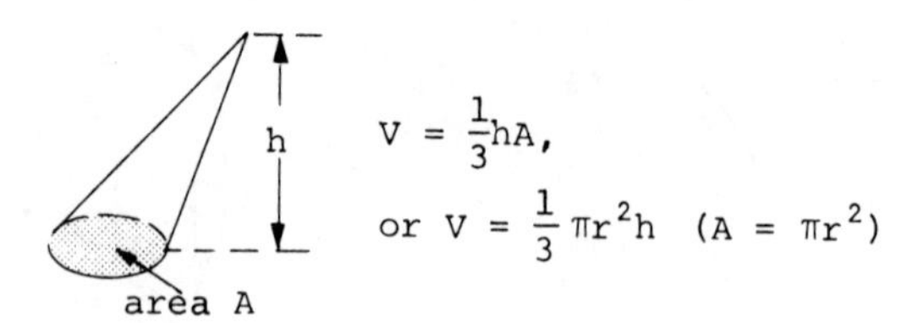

Theorem 8

The volume of a pyramid is equal to one-third the product of the area of its base and the measure of its altitude.

$$V = \frac{1}{3} Ah$$